Starchild

The Human Meanings
of the Big Bang Cosmos

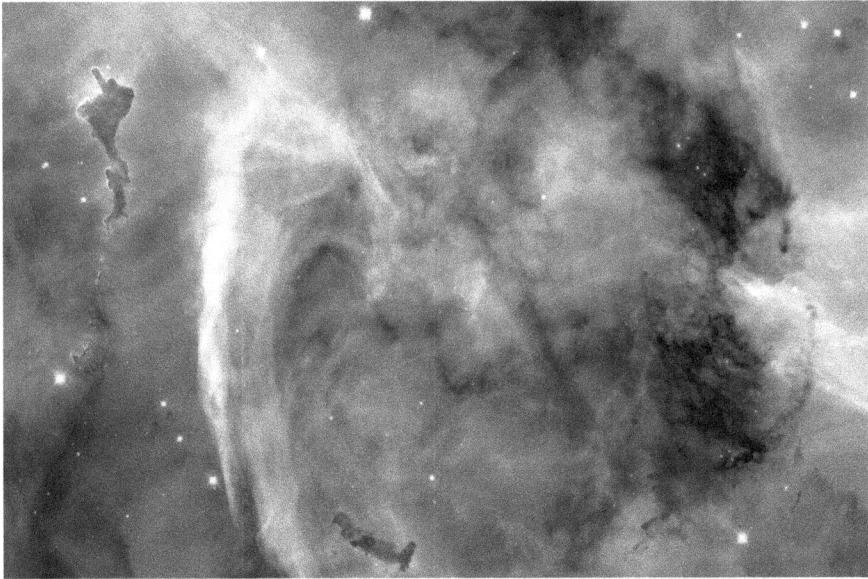

Don Lago

Plain View Press
P. O. 42255
Austin, TX 78704

plainviewpress.net
sb@plainviewpress.net
512-441-2452

Cover art photo courtesy of NASA, 2009:
Light and Shadow in the Carina Nebula

Cover design by Susan Bright

Acknowledgements

Grateful acknowledgement to the magazines in which these essays originally appeared:

"The Blizzard," "Fireflies," "The Spelling Lesson," "In the Sky with Diamonds," "You Can't Go Home Again," and "The Homing Instinct" in *Astronomy*; "Channeling the Big Bang," "Celebrating the Big Bang," and "Helios" in *Sky & Telescope*; "The Music of the Spheres" in *Isotope: A Journal of Literary Nature and Science Writing*; "The Particle Accelerator," and "Impression: Sunrise" in *The Trumpeter*; "Ripples" in *Balcones Review*; and "The Big Bang Discovers Itself" in *Pulpsmith*.

A special thanks to Susan Bright for believing in this book; to Richard Berry of *Astronomy* magazine for having the nerve to publish innovative literary writing in a science magazine; and to Caroline Werkley and Terra Waters, the best audience a writer could wish for.

Contents

Introduction: The New Creation Story   5

The Blizzard   11
Fireflies   15
The Particle Accelerator   19
The Spelling Lesson   29
In the Sky with Diamonds   35
The Hubble in the Bubble   41
The Hubble House   53
Lost in the Milky Way   63
You Can't Go Home Again   77
The Music of the Spheres   81
A Streetcar Named...   91
Through the Looking Glass   103
The Homing Instinct   107
The Big Bang Discovers Itself   113
Channeling the Big Bang   117
Ripples   119
Celebrating the Big Bang   123
Helios   129
A Birthday Wish   131
Impression: Sunrise   133

Endnotes   143
About the Author   145

Introduction: The New Creation Story

For as long as there were eyes on Earth, the stars gazed into them and asked a question.

For a long time the eyes of Earth gazed back without recognition. But at long last, one animal began to see the mystery of creation.

All over Earth, around campfires that signaled that the mad fires of stars and volcanoes and lightning had now been channeled through cells into a fire that nourished the building of order, human eyes flickered with starlight, and with a hunger that could not be satisfied by food. This was the hunger of unknowing. Humans yearned to understand the origin of their lives. For a sky and an Earth full of order and beauty, for the gift of life and for a life full of gifts, humans wanted to know who or what to thank. For a sky and an Earth full of chaos and dangers and suffering, humans wanted to know who or what to blame. To fill their hunger of unknowing, humans began to create creation stories.

With brains that began as heliotropic cells that could barely distinguish between patches of light and darkness, meaning-tropic humans sought the ultimate patterns behind the lightswarm cosmos. Humans recognized that their lives were part of patterns much larger than themselves. Groping to find patterns in the sky, humans saw the shapes of bears, birds, and fish, and then they reached beyond those visual patterns and tried to see patterns of events, to see purposes and causes and consequences, to see stories that could make sense out of a confusing world. In every society, small or large, humans created creation stories, thousands of creation stories.

In some stories there was nothing at all until a god uttered a thought or a word or a breath that became the universe. In other stories there was already chaos, formless matter that a god shaped into a cosmos. There were goddesses who gave birth to the cosmos, and giant eggs that hatched. There were gods whose semen or blood or sweat became the world. There were monsters that had to be slain and dismembered for their bodies to become the world. There were animals that became humans, and animals whose actions gave rise to humans. There were humans living in underworlds, from which they had to emerge. There were oceans into which an animal had to dive to emerge with earth. For every basic theme, there were numerous variations.

If creation stories were diverse, it reflected the diversity of Earth's landscapes and human lifeways. Mountains, river valleys, deserts, and Pacific islands left different imprints upon creation stories. Hunting, farming, fishing, and herding defined nature's gifts and challenges and human survival in different ways. The human imagination too was diverse, elaborating the same raw material into unique stories. Here and there a Stone Age Milton spun especially rich stories.

Yet the diversity of creation stories was not as significant as what they had in common. All over Earth, in every environment, humans spun stories that had the same psychological structure and purpose. Creation stories were the foundation of human identity. To define who we are, humans need to begin with where we came from and why we were created. Creation stories were usually the most sacred part of people's lives, the center of religious beliefs and rituals, of religious architecture and art and poetry. Creation stories usually incorporated the 'science' of the times, explanations for why nature works the way it does, yet science was not their primary purpose. Their purpose was to give structure to human consciousness, to assign value to human life, to give meaning and beauty to the world, to define relationships between humans and the rest of creation, to encourage or discourage various kinds of behavior, to offer support in the daily struggle for survival and sanity, to ease frustration and suffering, to deny that human efforts would always be defeated by death, and to affirm that life was worth all its troubles. As human societies evolved so did religious needs, and creation stories evolved from tribe-centered to universal, from nature-centered to society-centered.

From the start, humans picked up a trail, a storyline, that was not of their own making. They noticed patterns in the night sky, in the seasons, in the growth of plants, in the behavior of animals, in the workings of their own bodies. It was the Greeks who were struck by the idea that such patterns told a story greater than anything human had imagined, a story in which nature was a genius, matter was dynamic and powerful, and the universe was very large and perhaps very old. Humans followed the trail of this story slowly, often with confusion, sometimes with reluctance when nature began to crowd the gods from their thrones and hiding places. At the least, humans couldn't ignore that their understanding of nature gave them better tools, harvests, and health. This storyline led onward, through many surprises and strange ideas, ideas far beyond human experience and imagination. The universe grew ever larger; the stars grew more numerous and powerful. The earth became Earth, a planet floating through space, a geological engine sprouting mountains and volcanoes and rivers like spring flowers. Crude matter became atoms swarming with order and energy and talent. The life of Earth became not just predators or companions or totems or resources for humans, but our creator, a force that had been flowing for billions of years, electric with creativity, constantly transforming itself into new forms, polishing itself into new senses, convoluting itself into new brains and new perceptions, convoluting itself into ourselves. In the history of Earth and life, humans had appeared only recently. Humans followed this storyline ever father back in time and space. They found a cosmos aglow with strange objects and energies, a cosmos so immense that billions of galaxies were almost invisible. They found that evolution was at work not just for life, but for Earth, the solar system, and the whole universe. The universe was growing and changing; the galaxies were racing outward, racing from a single event, a moment of creation.

Without even intending to, humans have created a bold new creation story. This story is quite different from previous stories. It was created not by campfire Miltons but by a huge network of detectives with sophisticated instruments. It is vastly more detailed than previous stories. Most significantly, it was created not to satisfy the needs of human identity, but in following wherever the lines of reality led.

The new creation story has cast deep new dimensions into the old creation stories. The star patterns of bears and birds have ballooned into patterns unbearable light-years large, patterns flying through space with wings launched from the Big Bang. The monoliths of Stonehenge, intended to track the cycles of the seasons, now show the geological fingerprints of much greater cycles of time. Sacred mountains that were the realm of gods have turned into the playthings of tectonic forces that move continents and lift mountain ranges enough to house hundreds of gods. A star painted onto a cliff at Chaco Canyon turns out to have been painted by a carbon hand that was painted into complexity inside ancient stars. The Stone Age campfire Milton was speaking through a brain that held fourteen billion years of epic stories. The outreached hand of God upon the ceiling of the Sistine Chapel proved that God evolved from primates grasping tools and tree limbs in a long-forgotten forest.

Human creation stories were, it turned out, the product of a creativity much greater than that of humans. The same creativity that had built galaxies and solar systems and snowflakes and cells and trees was now building Stonehenges and pyramids and cathedrals and astronomical observatories to map out and affirm the order of the cosmos. It was the void of space itself that had knotted itself into the human brain and become the void of unknowing that tried to fill itself with myth and science. It was the universe itself recognizing its own long journey at last.

The new creation story has developed quickly by the pace of human culture, within a few generations, and it has offered such dramatic contrasts with previous stories that it has outpaced the ability of human culture to absorb it. The new creation story remains largely unincorporated into human identity. It remains abstract facts, not personal, lived realities. Humans haven't even caught up with the Copernican revolution; we still perceive sunsets and sunrises and not our planet turning and racing through space; even fewer people step outside into the night and perceive an expanding universe. Few people look at other forms of life and perceive all of us to be but shimmers of the same great river of life that has been flowing for billions of years. People may have heard the idea that the matter that composes us was crafted in stars and volcanoes, but few people perceive in their own heartbeats the echoing hammer strokes of stars and volcanoes. While previous creation stories have inspired the greatest poetry and art, the symbols by which ideas become lived realities, the new creation story remains largely invisible in poetry and art. This is partly because the scientists who have developed the new story are usually data-oriented and often lack the creative skills to translate data into dreams. It's also because the creative, humanities culture of the twentieth century has been heavily preoccupied with historic events and social realities. When the humanities culture has paid attention to the new creation story, it has often been hostile to it and eager to sound a retreat to the turtleback shamanistic cosmos of noble savages.

The largest reason why humans have been slow to translate the new creation story into human identity is that the new story is the first creation story not deliberately constructed to fit and support human identity. The new story is so different that it is not even clear how it fits human psychology. The first task of creation stories has always been to assign value to human life—and to human death—but the new creation story is lacking the usual architecture of value. At first glance many people see only what the new creation story loses; they see humans cast out onto a vast, alien, lonely, inhuman cosmic landscape. The new creation story does indeed involve losses, some serious losses. But it also offers many rich new possibilities. It offers a different kind of magic than the shamanistic magic of chanting a seed into a cornstalk; it offers the magic of a seed so ingenious and powerful that it grows into a cosmos echoing with chants of gratitude for being alive. The new story may make human bodies small in size but it fills those bodies with a whole universe of time and evolution and ability. Not all of the losses of the new creation story are bad losses. Previous creation stories have happily served the absurd narcissism of humans, a narcissism that has become incompatible with human survival in an age of nuclear weapons and global environmental threats. A large dose of humility, of recognizing that we belong to life and the earth and not they to us, may be essential to preventing humans from ruining eons of creation.

"There is a grandeur in this view of life," wrote Charles Darwin, coaxing us to refocus our vision. There is grandeur in a cosmos that is far more ingenious and powerful at unfolding order and energy and life than were any of the old creation stories. We have barely begun to explore the new creation story and its meanings for humans. This should be the most important cultural task of our era.

This book explores the new creation story—the new cosmos—and the place of humans within it. I seek to translate science into the symbols, the metaphors, the poetry that catalyze the human imagination, and that turn outer realities into inner realities, facts into identity, ideas into emotions, stories about life into life itself. I use the keys of creative nonfiction, of literary nature writing, to try to unlock the meanings, the beauty, the grandeur of the new creation story. I explore the Big Bang cosmos, the integral place of life within it, the men and women who discovered it, the process of discovery, and the ways we might celebrate the new story. I do not ignore the losses of the new creation story, the loss of values. But the new creation story does offer its own answer to questions of value. The universe "valued" life right from the start. The energies of the Big Bang were instantly generating order; the structures of the atom supported the generation of further order and of life; and cells devotedly wove the order of bodies until those bodies consciously felt that living was important and wove creation myths to justify why. This 'why' may now include the answer that the life of Earth is a masterpiece of a very long, hard, creative work, incredibly rare and precious, calling for celebration.

The Blizzard

How can a human grasp the scale of the universe? The scale of the universe is vastly beyond human experience. Tonight my experience of the universe has consisted of walking two or three miles down an old trail to get away from city lights. Two or three miles can seem like a long way for a human. I find that distances seem longer when I am walking at night. Perhaps this is because I am moving slower to avoid tripping on something, or perhaps it's because the usual visual cues about my rate of speed are gone, or perhaps, in some relativity effect, the universe actually gets larger at night. I am tempted to believe this latter theory. By day the entire universe may consist of an office, a classroom, a car, or at most a blue sky, but at sunset the claustrophobic roof of our lives is rolled away and we are invited to see the real dimensions of where we live. The universe also becomes more mysterious at night, if only because humans are visual animals and at night our familiar landmarks are hidden or blurred. But it's also because the night sky really is mysterious.

Two or three miles did register on my human scale of distance, and it made a big difference in the geography of human suburbs and streetlights, but now that I had left town for astronomical realms, I realized that my mileage didn't amount to anything. At this rate I would take about ten years just to walk to the moon. My feeble mathematical skills didn't even want to deal with how many millions of human lifetimes it would take to walk to another star.

Cosmic distances are only the beginning of the unscaleable scale of the universe. There's also the size of astronomical objects. There's the scale of time billions of years deep. There's the scale and strangeness of the energies being fired from stars and dead stars. There's the depth of creativity that enabled an outburst of energy, the Big Bang, to evolve into a cosmos of order and life.

My astronomical self-assignment for tonight was to make a modest attempt to calibrate the scale of the universe. In my backpack I carried binoculars that should be good for observing some galaxies. Galaxies are good for getting a feel for cosmic distances. You see a tiny patch of light yet you know that it contains billions of stars. Around those stars flow planets and comets and asteroids and moons, and on those planets and moons are billions of oceans and mountain ranges and canyons and craters. Some of those planets are effervescent with life, including intelligent life that is standing in the night and gazing at a tiny patch of light that is our galaxy. Seeing so much reduced to so little might prompt me into recognizing the vast distances of space. Perhaps if I stared at a galaxy long enough and tried really hard to believe how large it was and how far away it was, I might almost succeed. It might offer a mirror's recognition that the stars scattered around me seemingly at random were actually part of a similar tiny disk of light surrounded by blackness. I might notice in that mirror how life was a tiny and rare thing in the universe. I might even see in that mirror something different about my own face.

The sky was already partly cloudy when I'd set out, but for a winter's night it wasn't unpleasantly cold. Yet as I walked, more clouds began rolling in, and by the time I reached some nice dark countryside, clouds stained half the sky.

I sat down on the edge of a meadow. I looked up and found a spot where a galaxy should be. I raised my binoculars and hunted for the galaxy, but saw only stars and darkness. Perhaps my binoculars weren't quite good enough, or perhaps the increasing haze was blurring my vision, or perhaps I wasn't even looking at the right spot. I began looking for another galaxy, but clouds soon covered that whole direction. Soon the sky was almost completely covered by thick, dark, pregnant clouds. I waited for the sky to clear. I waited ten minutes, fifteen, more, but now not a patch of clear sky remained, and the breeze blew harder and colder. At last I realized I wasn't going to be seeing any galaxies tonight, and I headed home.

As I walked toward town, my mind full of galaxies and disappointment at not seeing any, a few snowflakes flickered past me. Snowflakes two or three at a time, then more, then many. The snowflakes turned the air into a fluttering whiteness. My face tingled with tiny cold touches. The snow grew thicker and thicker. Whiteness closed in around me. The shapes of trees grew vague and then almost disappeared in the whiteness. Distant streetlights turned into ghosts. In the shifting breeze the falling snowflakes gently flowed this way and that; I was passing through one billowing white curtain after another.

At first I noticed the snowfall only as an adversary that had ruined my little astronomy expedition. But slowly the snowfall penetrated my mind, and I began to see and enjoy it—this ticker-tape parade the universe throws for its creatures. I looked upward and watched snowflakes emerging from the vague whiteness and drifting down. I watched them coming and coming without pause, flowing past me by the hundreds every moment, by the thousands. And then I saw something more than just a lovely ticker-tape parade. The snowflakes were just like—galaxies.

The tiny white disk of a snowflake was what a galaxy would look like from afar. It was just like what I would have seen through my binoculars. The flow of thousands of tiny white disks was what you would see if you could sit stationary in space for a billion years and watch galaxies flowing away from the Big Bang. I stopped and stood there watching the flow of tiny white disks, and now I was seeing galaxies flowing through space. I was seeing the universe's creativity, its ability to create order, for these galaxies had crystallized out of formlessness, and each one was unique. As the galaxies flowed and flowed without pause, I was seeing billions of galaxies; I was seeing the massive prodigality of the universe. I was seeing massive power, for each of these disks contained billions of stars furiously pouring out light and heat; yet I was also seeing massive emptiness, for space is so vast that all of that light and heat is reduced to a cold flake. I knew that hidden within each of these disks were billions of planets and moons covered with oceans and glaciers, mountains and canyons, deserts and craters. I knew that within these disks were multitudes of life, and that one single disk might hold a million civilizations. I watched the galaxies flowing and flowing and endlessly flowing, flowing from the Big Bang outward. And then among a thousand galaxies I spotted our own, the Milky Way, and I watched it flying and spinning through the darkness, so tiny and frail; I saw within it our Earth, and I saw myself standing in a snowstorm, which was a tiny local manifestation of the cosmic blizzard.

As the cosmic blizzard flowed around me, I saw the scale and the power and the grandeur and the beauty of the universe as I had never seen it before. The blizzard blurred the familiar outlines of the world and of myself, and prompted me to see the world's mystery. I walked home with the opposite of disappointment; I had wanted to see galaxies, and galaxies were what I saw.

Fireflies

As the horizon covers the sun and the daylight fades into twilight and finally into night, darkness seems to take over the world. Yet this ending of daylight is also a beginning. For the stars, dusk is dawn. More distant suns appear and shine their fainter daylight onto Earth. Good morning, starshine. And good morning, galaxyshine. As lights emerge in the sky they are joined by human-made lights on the ground; a city turns into a galaxy complete with spiral arms and wandering stars.

Tonight I was walking away from the human-made stars so that I might enjoy the real stars. I headed down a country road, turned into some woods, crossed a stream, and emerged in a large meadow of tall prairie grasses, a meadow circled by trees. I stood there and looked at the woods silhouetted against the sky, a sky full of stars twinkling. The same air that made ripples in the starlight now moved over the land and made ripples in the grass.

As I approached the trees, I saw that they too seemed to be full of stars. Yet these stars were flashing on and off, and some of them were moving about. It was June, and the fireflies were mating. It was the Midwest, and the air was thick with moisture, warmth, and insects. The trees contained hundreds of fireflies, pale green lights that glowed for several seconds and faded off and then glowed again. The prairie grass too held lots of lights, lights blinking on and off. As I looked at the lights closest to me, I saw that they illuminated vaguely the stalks and leaves and flowers on which they sat. Now and then a light detached itself from the grass and climbed into the air, and others broke from the trees and drifted toward the ground like autumn leaves. As the fireflies moved through the air they traced out their pathway for awhile, then disappeared, and then showed themselves again a short distance away. Often the flying lights curved this way and that, spelling out a 'J' or 'S' that may have had more meaning to other fireflies than to me.

I stood there and watched the crowd of lights shifting about and flickering on and off, while above them the unmoving stars flickered more steadily. I enjoyed the way the lights in the trees blended with the lights in the sky. With a little imagination I saw the fireflies as stars, and saw myself amid a strange constellation, or in a planetarium where stellar motions are speeded up. And then I realized that the fireflies truly were stars; or rather, they were all pieces of one star, the sun.

The light being emitted by the fireflies had once been emitted by the sun. This energy had been squeezed from matter in the heart of the sun, and it had flowed in dense, turbulent currents of matter and energy.

Gradually it worked its way outward, and finally it reached the surface and flew into space. It traveled across millions of miles. Then it ran into a planet of blue and white and green. It streaked through the blue, passed the white, and fell onto the green. It fell onto leaves and was snared in tiny chemical traps. The energy was given a collar and led about and put to work. It was made to bind molecules together or to break molecules apart. It was made to energize all kinds of processes. These little pieces of the sun flowed through cells full of activities quite unlike the activities they had known inside the sun. Then one day some fireflies came along and nibbled on these plants, swallowing mouthfuls of sunlight. The sunlight flowed through the chambers and passageways of the fireflies. As it flowed into one chamber it was touched by one type of molecule and set loose to fly through the translucent skin of the fireflies. After a long imprisonment, the sunlight was flying free again.

The lights before me were tiny pieces of vanished days, days the fireflies had revived for one last flicker. As the fireflies flashed I saw little pieces of last Thursday or last Tuesday or last Sunday; I saw days of weeks or months ago. Most likely, each flash was a mixture of many days. I watched the yesterdays flashing like memories. I thought of how the sunlight of those days had warmed my skin and guided my journeys and grown my vegetables and carried the shapes of letters from books to my eyes. When those days had come to dusk I had thought they were over, but here they were again, shining on me once again. The yesterdays flashed into space, and tried to catch up with their fellow light from those days.

Yet something about this light was very different now. The light that had shined on me last week had been produced by brute force, by massive amounts of matter forced by gravity to pound against itself and rip itself apart. The light shining on me now was generated by something refined and skilled; in the fireflies matter had learned the secrets of its own workings and codified that knowledge into genes and into the molecular keys that could gently unlock light from matter. In the sun the light had been helplessly swept along on waves of matter and energy and spewed into space like smoke from an engine, but in the fireflies light had become the driving force for life. In the sun the light had contributed to the chaos of the sun, but in the fireflies it helped build a highly intricate order. In the sun the light had constantly collided with other particles and ricocheted away, but in the fireflies light tied together long and complicated molecules. From the sun the light had been fired aimlessly into space, meant for no one, but for the fireflies this light was a signal full of information for other fireflies. Most of the light that poured from the sun disappeared futilely into space, but the light of the fireflies would lead to the merging of genes and the creation of new life.

As I watched the fireflies flashing and lifting off and swirling against the light of the stars, I saw that the wild fierce energies of the universe, the energy pouring massively from countless stars, had crafted itself into the gentle energies of life. I saw that the ceaseless streaming of matter—the streaming of electrons in atoms, the swarming of particles inside stars, the swirling of moons around planets and of planets around stars and of stars around one another and around the galaxy, and the streaming of galaxies away from the Big Bang—this wild fierce dance of matter had choreographed itself into the gentle dance of molecules through cells, which multiplies itself into all the forms and activities of life. As I looked at the trees and the grass before me I saw them as a skillful dancing of particles, and I saw this dancing of particles multiplied into the mating dance of fireflies. The fireflies flickered against the stars, and I saw that the blindly rushing matter and energy of the universe had taken on wisdom and purpose. In the fireflies, starlight had become the light of consciousness. And in the midst of the stars and trees and grass and fireflies I saw myself, a human form—the blind furious cosmic dance turned into a mind that could understand it all.

As I watched the fireflies flickering on and off, on and off, a new vision suggested itself and began to grow.

I was looking at the contrast between the blinking fireflies and the steady light of the stars when I realized that the stars were not really steady. In a longer perspective, the stars too blinked on and then off. The stars would blink on and off only once, and they would stay on for millions or billions of years. With a little imagination, I saw the fireflies as stars forming from gas and igniting and drifting through space for awhile, and then dying out. I watched a light blink on, and around this light I imagined planets forming and evolving in various ways. I watched the star wander among other wandering stars. Finally I saw the star exhaust its fuel and turn dark to wander invisibly through darkness. One after another stars blinked on: I witnessed the prodigality of the universe. Yet as the stars disappeared just as steadily I also saw how fleeting even stars could be in this universe.

Then I was seeing the fireflies not as stars but as whole galaxies. Each light blinking on was a galaxy forming out of the formlessness of the Big Bang. Each light blinking on was billions of stars igniting, stars that swirled around the galactic nucleus, stars that continued turning on and off by the millions but that in their combined masses maintained a steady glow. I watched galaxies flying through space. I saw them swerve and cluster. And then I saw the galaxies turning off.

Finally I saw a firefly as the entire universe.

A firefly blinked on, and I was seeing the Big Bang itself. The firefly glowed for a moment, and within this glow I saw billions of galaxies glowing. Then the glow faded away, for all the galaxies were dying. Finally there was only darkness—only darkened galaxies rolling through the darkness forever.

When seen from the perspective of eternity, this is what the universe will seem like—a firefly's flash. The massive light of the Big Bang, the fierce outpourings of quasars and supernovae, the outpourings of all the stars of billions of galaxies, all are contained within this flash of light. This flash also contains the entire history of life and civilizations: the long groping evolution of chemicals into cells and into larger forms and into intelligence; the creation of tools and languages and societies and science; the exploration of solar systems and the spread of civilizations from star to star. Perhaps billions of civilizations will cover their planets with machines that pour out energy, yet all of this energy will be only a tiny part of this flash of energy. On perhaps billions of planets fireflies will flash in the night, yet this light too will be but a tiny part of a much larger firefly's flash.

There is something in humans that feels threatened by such a vision. Even the prospect of individual death can make us feel insignificant, can make us cling to things through which we can live on: families, organizations, memories, tombstones. The prospect of the entire universe dying out into eternal darkness doesn't mesh well with human psychology. Yet perhaps this only proves that something is wrong with human psychology.

It is significant that the universe contained any light at all, for the universe might have been so crude that it never became anything more than dark gas. It is even more significant that the universe created light for many tens of billions of years. It is significant that some of that light turned into life, that it flowed into bodies and inspired molecules to dance and weave themselves into intricate patterns, that it lifted dust and water out of the ground and turned them into forests and meadows, that it became the soft green glow of fireflies, that it became the pulses that kept uncountable hearts beating steadily, that it carried images from eyes to brains, that it powered the walking of animals, the swimming of fish, and the soaring of birds on billions of planets. It is significant that some of that light flowed into brains and transformed itself into a brighter kind of light, a light that enables the universe to see itself. Instead of a universe forever dark and unknowing, the universe has shined this light upon itself and its brilliant journey.

The Particle Accelerator

The wind was flowing to the north, and so was spring.

Out of its winter hiding places, life was emerging. The land was turning from black and white into color. Mammals, reptiles, and insects had unthawed from a winter's sleep and were moving about. The air was filling with seeds searching for places to bloom. These seeds were not just those sprayed from plants, but also those riding inside the bodies of birds.

Upon the northward wind a flock of swans was heading for their breeding ground in the north of Canada. Feeling the same primordial force that stirred every form of life into its own version of spring, the swans had left their winter home on Chesapeake Bay. They had risen over the Appalachian Mountains and swung along the southern edge of the Great Lakes. They were heading almost to Alaska in a journey of some two thousand miles. Right now the swans were crossing the farmland at the tip of Lake Michigan. The sunset was coloring the sky and the clouds, yet for the swans this was still early in another day's flight.

With their long necks stretched out straight before them, with their legs tucked carefully against their tail for streamlining, the forty swans chopped the air with their wings, with a blending of power and grace. Except for their beaks and feet, the swans were as white as if they were clouds that had condensed right through the state of water and into the state of life.

The light faded quickly when the swans flew into the brewing clouds of a line of thunderstorms. Flying at several thousand feet, the swans were caught in the powerful currents inside the storm clouds, which heaved them up and down, jerked them about, and skewed their formation. Rain pounded against them, and thunder roared through them. Calling out to one another in their high-pitched voices, the swans struggled to stay together and to escape the winds. They descended rapidly through the storm. When they were close enough to the ground to see through the rain, they searched for a place to land. Soon they saw what appeared to be a giant circle on the ground, and amid the land inside it they saw a lake.

The swans skimmed over the trees beside the lake and descended, holding their wings to brake. As they neared the water they lowered their feet and spread wide their webbed toes. They swung their bodies vertical, pointing their tails downward and their feet forward. As they plunged their feet into the water, the energy accumulated by forty pairs of beating wings was translated into forty large wakes spreading behind the swans.

In the control room at the edge of the particle accelerator, which formed a circle four miles round in the Illinois prairie, physicists prepared for another probe into the secrets of the universe. The Fermilab accelerator, one of the world's most powerful, would generate a stream of particles, boost them to enormous speed and energy, and then aim them at a metal plate. The energy of impact would shatter the target atoms into their constituent particles, and create new particles. These particles would spray through a bubble chamber, which contained a special liquid that would turn into a trail of bubbles where a particle passed through. From these wakes the scientists could determine a particle's size, speed, charge, and spin, and perhaps glimpse a little deeper into the structure of matter.

The scientists moved down the long control panels, pushing switches and checking gauges and computer screens to bring every part of the complex system into readiness. They readied the power generators to fire off the equivalent of many lightning bolts. They readied electrodes to apply that energy onto the particles. They activated long chains of powerful magnets to guide the particles in flight. They made sure that the vacuum pipes were empty and the superconductors -450° F cold. They brought the network of sensors and computers to life, and told them how to run this experiment. They checked that the target was in place and that the bubble chamber, cameras, particle counters, and other devices were ready to detect the results. When the scientists saw that each of hundreds of systems and subsystems was set to precision, ready to work in microsecond synchronization with one another, the director gave the signal to begin.

In the lake in the middle of the particle accelerator, the swans glided across the water, drawing wakes upon it, wakes that overrode the wide ripples drawn by the wind and the tiny circular ripples from the thousands of raindrops striking the water every moment. In the distance, clouds struck the ground more powerfully with a lightning bolt, sending equally powerful ripples spreading through the air for miles. The swans moved toward the shore and drifted along it until they found a shallow cove that offered both shelter and food. While they waited for the storm to pass, the swans began storing up more energy to turn into wind.

The swans poked their heads into the water and tugged small plants out of the mud, or skimmed their beaks along the water's surface to collect algae. They raised their necks and swallowed. They curled their necks down again and dipped their beaks to suck in a mouthful of water, a few atoms of which had just been part of the huge, turbulent power of the thunderstorm and swirled down from the sky. Now the water swirled down into a more sophisticated flow.

Power surged from the Cockcroft Walton generator and set flowing a stream of hydrogen atoms, the basic atom of the universe, the atoms forged in the first moments of the Big Bang, the atoms from which all other matter has been built in the course of cosmic evolution. To the single proton and electron of the hydrogen atom, the generator added another electron to make the atoms electrically charged so that magnetic fields could grip them. Boosting the atoms to an energy of 750,000 electron volts, the generator fired them into the next stage of the process, the 50-foot-long and straight Linear Accelerator.

The atoms flowed through a narrow vacuum pipe surrounded by huge electromagnets, which kept the atoms focused in the middle of the pipe. Every foot or so an electrode poured energy into the vacuum pipe, setting up an electrical surf wave on which the atoms rode, giving the atoms speed and energy, up to 200 million electron volts. In only an instant the atoms had raced through this throat and into the third of five stages of acceleration.

The water molecules, which were made of two hydrogen atoms and one oxygen atom, trickled down the throat of a swan and into the stomach. The water flowed into a soup of algae, leaves, stems, and roots, which were slowly being dissolved so that their matter and energy could be absorbed into the swan.

These plants were packed with photons that recently had arrived from the sun, where the photons had taken part in pushing hydrogen atoms up the first step into higher order. From that hydrogen storm the photons had come to a world where matter had journeyed far indeed into higher order, and now the photons were absorbed into that order to help push it onward. The plants also contained particles absorbed from the earth, particles that had been imprisoned in rock for a billion years and then been torn loose to join other crude masses like mountains and deserts for more ages, and to be volleyed in the wind and swept down a thousand rivers to the sea. The plants also contained water particles that had flowed turbulently in rivers and oceans for ages. Together the crudeness of sun, land, and water had been siphoned into swans' stomachs, and now it was being molded into the shape of a swan.

Many miles away, leaves fell into another kind of stomach. Time and the earth had changed these leaves from soft green into hard black, but they were still packed with particles they had absorbed from the sun. In a metal stomach the leaves were digested by fire, and they released their sunlight. Energy that might have powered the flight of pterodactyls now raced through power lines, raced across the countryside, raced into the Booster Accelerator, a circular machine 500 feet in diameter, where it stripped the racing atoms of their electrons and swept the remaining protons around and around, transferring to them an energy of eight billion electron volts. Pushed by the combined effort of trillions of ancient cells, the protons raced into the Main Accelerator.

Tied inside molecules of water and food, the protons flowed through the pores of the stomach and intestines and into the bloodstream. They were swept through the tree of veins and into the heart, where a steady electrical pulse was magnified into a steady pulsing of flesh. The heart sent the protons racing throughout the body. They swept into the lungs and helped capture a tiny portion of Earth's crudely surging air to make surge the weather of bodies. They flowed into cells and joined the waltz of molecules, helping to carry giant molecules back and forth, helping to break them apart and bind them together, helping DNA unfurl and copy itself, helping cells grow and split into more cells. The protons joined dozens of types of molecules performing dozens of types of tasks in dozens of types of cells. They became the firmness of bone and the softness of eyes. They became the power and beauty of a swan's wings. And when the protons joined the pulsing energies and particles of the brain, they looked out—through eyes where light caught by leaves now energized the catching of light in optical leaves—and saw the beauty of a swan's wings, saw forty swans gliding together on a lake of silver energy, saw the surface of the lake rippling to the breeze and the raindrops and the light of the disappearing sun, saw the prairie and trees and human buildings around the lake, and saw the clouds beginning to part and reveal the deep lights and spaces of the universe.

The Main Accelerator applied far more power to the protons, sweeping them around the four-mile underground track until they approached the speed of light and were circling some 50,000 times every second. At a precise microsecond in each lap, the accelerator poured a wave of energy into the slice of outer space inside the vacuum pipe, and it also poured energy through the surrounding magnets to keep the protons focused in the center of the pipe and to turn them as the pipe turned. The Main Accelerator contained two layers one atop the other, and when the top layer had accelerated the protons to 150 billion electron volts, it fired them into the lower layer.

When the rain stopped, the swans shook themselves vigorously to throw off water and to fluff their feathers. Accelerating air through their long necks to call out to one another, they drifted into the center of the lake to give themselves a long runway for accelerating.

The swans began paddling hard with their feet and surged forward. They spread their wings and waved, waved once a second and then faster, lifting themselves up until they were running upon the surface of the water, their feet slapping it with force enough to raise splashes a foot high. The energy of each rowing of wings rippled through their bodies and caused their tails to flap and their outstretched necks to undulate. Side by side, half a wingspan between them, the swans raced across the water, and then they were free of it and slowly gaining altitude and speed.

The swans crossed the shoreline, crystallized into a V formation, and began swinging in a large circle. They flew a circle four miles long around the lake, sensing out the direction and strength of Earth's magnetic field. Inside them, particles swarmed to calculate the direction home, the direction to that one lakeshore where the swans were born. Protons swarmed to accelerate themselves to home.

Speeding directly beneath the circling swans were other protons, and they too were migrating back to a place they hadn't been in a long time. Theirs was a much vaster migration, reaching across many light-years and across fourteen billion years of time to where the protons were born, to where the cosmic egg had cracked open with a big bang and the universe emerged.

In the course of evolution on Earth, the force that began as a feeble flowing of particles inside a cell had grown increasingly sophisticated and powerful. Through leaves life had tied its flow to the powerful flow inside the sun. Cells expanded into many-celled bodies, with particles, energy, and information flowing throughout those bodies and between them. Bodies developed ingenious ways of accelerating themselves and moving about. In brains this flowing quickened itself into knowing, until humans were wondering what all the flowings of Earth and sky were made of and where they had come from. Humans built huge instruments to find out. With this particle accelerator, this circle like a hugely magnified cell with a lake as nucleus, the cells of Earth had hugely magnified their skill at moving particles, magnified it into the power and precision necessary to boost particles into a re-creation of the Big Bang.

The protons raced into the lower ring of the Main Accelerator, into the powerful electromagnetic grip of a chain of one thousand superconducting magnets, which a vast refrigeration system worked to keep at a temperature as cold as space so that electric currents could flow freely through them. The protons were pushed even faster by continuing pulses of energy, pulses that were really an amplification of human brainwaves, the brainwaves that formed ultimate questions and conceived machines to answer them. From these waves the protons gained an energy of one trillion electron volts. Then the magnets snapped the protons out of their circle and raced them down a long straight track, at the end of which all the energy of their motion was instantly turned into an energy of impact that matter hadn't reached since the first microsecond of the Big Bang. The protons smashed into the target atoms and overwhelmed the energy that held the atoms together, sending their pieces spraying apart to swerve and spin according to their varied natures. This energy of impact was so intense that it also transformed itself into new particles too energetic to exist in normal conditions, particles which flashed through the bubble chamber for an instant and disintegrated into other particles and back into energy.

In only an instant this explosion, this glimmer of the Big Bang, was over. Compared to the real Big Bang, the scale of this explosion was tiny indeed. Yet when the products of this collision and thousands like it flowed through the bubble chamber inside the human skull, these collisions would create again the whole Big Bang and all the matter and energy in the universe.

As the swans broke from their circle and headed north, they flew directly over the time warp to the Big Bang, over the particles speeding and colliding in the way every particle in their bodies had done long ago. While the matter in the accelerator was being stripped naked to reveal its deepest structure, its most basic bonds and forces, the swans were clothed in fourteen billion years of order and ability. Atop the gluons binding the swans together, they were also bound by love. Atop the physical constants that gave all their particles a similar form, they now carried DNA that gave them the similar form of a swan. Atop the electromagnetic force that kept their electrons flowing, they now flowed out of a fierce determination to turn on their DNA and let the swans coiled within it materialize. Like particles in the Big Bang, the swans' acceleration would lead to their splitting apart, but the swans would split more gently by squeezing eggs out of their bodies.

The swans climbed hundreds of feet and might have risen for thousands if not for the clouds. Steadily waving their wings, they cruised about thirty miles an hour, surging when the wind surged.

As they moved north, the swans passed over land they had last seen in the autumn. Then the land had been dying, the grasses shriveling up and the trees turning from green to red as they bled their life away into skeletons. Insects, reptiles, birds, and mammals had been disappearing into sleep, migration, or death.

Now this land was erupting with life. In the prairies, in the woods, in the farmers' fields, in the city parks and backyard gardens, in the waters of lakes and rivers, seeds were switching on and plants were sucking in soil, water, air, and sun and spinning them into cells. Plants of thousands of shapes fit together like jigsaw pieces to form a solid green, a Mondrian picture of the vigor of life. Here and there amid the green, the neon signs of flowers advertised their particular forms of life. Plants were being transformed into the tens of thousands of kinds of insects swarming among them. Everywhere eggs were hatching. Animals were accelerating on the ground, underground, and through the trees. The streams and lakes were flowing into the veins of fish and, with fins, signing their new name upon the water's scribbling surface. In the air beneath them and alongside them the swans passed many other birds, some of them also in large formations migrating homeward, all of them inhaling the wind to make it blow them along from within. On their northward journey the swans were seeing a tide of life sweeping across a whole continent.

As the swans flew on, they left the clouds behind and entered a sky of thousands of lights. Among those lights were the other planets of the solar system, none of which were anything like Earth. As large as the life of Earth might seem to the swans as they flew over it, that life was very small and rare in the universe. In the vastness of space, all the stars and planets were but tiny particles, among which one blue particle's surface had begun flowing in tiny but rich patterns.

The swans flew on through the night, now using the constellations for direction, calling to one another to bond their own living constellation.

Among those thousands of stars there was one light, near the horizon on the swans' right, that humans officially called NGC 6618 and informally called the Swan Nebula. The nebula's shape, a white tapering body with a stubby neck rising from one end and expanding into a head and beak, even with a dark patch for an eye, reminded humans of a swan at rest.

Or perhaps a swan sitting on a nest.

The Swan Nebula was a huge cloud of gas and dust condensing to form stars. Perhaps the blast wave from a supernova had swept through space like a warm front and started the gas and dust condensing, or perhaps the weather of space had more gradually gathered matter until it condensed upon its own gravity. Now the nebula was brewing like a thundercloud, full of massive currents of matter and energy. Some of those currents were spiraling inward and thickening into stars. Where stars had ignited, they poured light through the cloud and made it glow. When enough stars ignited, their out-rushing light would stop the nebula from condensing further and begin to disperse it.

When the swan drifted away, she would leave a batch of eggs, the planets solidified from her body. For ages those planets would incubate in the warmth of their stars, and one day one of them might hatch. From that egg would fly millions of swans, their white feathers a condensation of that white nebula, their convoluted brains a knot tied from those clouds. From that egg would fly, swim, crawl, hop, walk, jump, run, float, tunnel, waddle, slide, and march innumerable offspring.

The swan offspring of the Swan Nebula would flow back and forth across their planet in seasonal migrations. Each spring they would feel a rising impulse to return to the place of their birth. They would cross bays and mountains, prairies and lakes, flying through storms and crystal skies, through day and night, guiding by scents and landscapes, magnetic fields and stars. At the end of their flight they would give birth to new swans, who would carry on that ancient migration.

Indeed, the Swan Nebula swans would have inherited this migration from their ultimate parent, the Swan Nebula itself. She too had completed a long migration to reach the place where she could lay her eggs. She had flown all the way across the universe. She had migrated from the Big Bang. This was a migration not just in distance, but also in form. The Swan Nebula had traveled through all the phases of cosmic evolution, from the making of quarks to the making of galaxies, through each level of particle building in the Big Bang and atom building in the stars. The Swan Nebula was still migrating, heading into the next phase of cosmic evolution, into the building of solar systems, geologies, and weathers. And then the Swan Nebula would flow into life.

The swans born from that nebula, and the swans and all the life hatched from the egg Earth, were the culmination of a great flowing of matter through fourteen billion years, a flowing begun by the Big Bang. The Big Bang sent stars and planets racing through space like protons in a particle accelerator. The Big Bang began this acceleration with real protons and other basic particles, but as they flowed onward they began flowing in more elaborate patterns and assuming more elaborate forms, becoming stars and planets, lakes and thunderstorms, leaves and the human brain. The Big Bang accelerated raw particles into the grace and love of migrating swans. The Big Bang accelerated particles until they formed an image of the Big Bang accelerating nothing into everything.

The Spelling Lesson

The clock said two o'clock, and Mrs. Rosetta told her class to get out their pencils and notebooks. It was time for the spelling lesson.

"Today we are going to learn words of science," she said. "Who can tell me what science is?"

Several children shot up their hands. Mrs. Rosetta called on Jennifer.

"Science is finding out things about the world," said Jennifer.

"That's a pretty good definition of science," said Mrs. Rosetta. She wrote "SCIENCE" on the blackboard, and the children copied it down in their notebooks. "Let's start with the word that includes everything else: Universe. The universe is everything there is—space, the stars, the planets, your houses, and you too." She wrote "UNIVERSE" on the blackboard and said, "Repeat after me: U.N.I.V.E.R.S.E."

The children said in unison: "U.N.I.V.E.R.S.E," and copied it down.

"Scientists think that the universe began very long ago. You remember we talked about this in science class. The universe began in a great explosion called the Big Bang." She wrote "BIG BANG" on the blackboard and the children copied it, except for Jimmy Lux, who leaned toward Adam Jones and whispered, "Wanna see a big bang?" From his mouth emerged a pink bubble, which quickly swelled—then popped.

Out of nothingness it emerged, a tiny bubble of space-time. It swelled with incredible speed, and the energy of its expansion was transformed into a deluge of radiation and particles, swarming madly. But at once the particles bound themselves into larger particles. Out of chaos, order began being born.

"The next word is quark: Q.U.A.R.K."

The class repeated, "Q.U.A.R.K." Several of the children giggled.

"Quark is sort of a funny word, isn't it?" said Mrs. Rosetta. "But it's an important thing. Quarks are little bitty particles, so small you can't see them. But they make up everything in the world: stars, rocks, trees, kittens, and even all of you are made of quarks. It's by arranging quarks in different ways that you can make so many different kinds of things."

As different types of quarks combined in different ways, they created many kinds of particles, including neutrons and protons, which began combining into atomic nuclei. When stars formed, neutrons and protons were fused into heavier nuclei, such as oxygen, calcium, and carbon.

"I'm sure you all know what a star is. Who knows how to spell 'star'?"

Several arms made of oxygen, calcium, and carbon shot into the air, fingers wiggling. Mrs. Rosetta called on Claire.

"Star is spelled S.T.A.R."

"That's right, said Mrs. Rosetta, writing it on the blackboard. "And a huge bunch of stars is called a galaxy." She wrote GALAXY on the blackboard, pronouncing each letter.

The children repeated, "G.A.L.A.X.Y."

"And what is it that moves around a star? Claire?"

"Planets," said Claire.

"Right. Do you know how to spell 'planet'?"

Claire tried each letter slowly: "P.L.E.N....P.L.A.N.I.T."

"Almost. Does anyone else want to try?"

Venus raised her hand. "P.L.A.N.E.T."

"Good, Venus." Mrs. Rosetta wrote it on the blackboard. "Everybody. P.L.A.N.E.T."

"P.L.A.N.E.T.," the children said and wrote down.

"And what goes around a planet? A moon. Can anyone spell moon for us?"

Venus tried "M.O.N.E.," and then Maria Luna got it right: "M.O.O.N."

"Good, Maria. M.O.O.N."

The class orbited: "M.O.O.N."

As order emerged on the atomic scale, it also emerged on the cosmic scale. The chaotic matter flowing out of the Big Bang began dividing into billions of clumps, into galaxies, which then divided into billions of stars. Around stars matter began forming a new kind of body. The conditions weren't always right, and often the effort failed. But here and there matter managed to form P.L.A.N.E.T.S. At first the only planets were made of gas—hydrogen and helium. Yet as stars created heavier elements and sprayed them into space, planets of solid matter formed. Different elements and different circumstances formed planets of ice or rock, oceans or deserts. Around planets there formed M.O.O.N.S.

"We've spelled a lot of words of astronomy," said Mrs. Rosetta. "Now let's try some words of biology. Biology is the science that studies life." She wrote LIFE on the blackboard. "Let's spell 'life'. L.I.F.E."

"L.I.F.E.," said two dozen lives.

On some planets a new kind of order emerged. Carbon atoms flowed through the oceans and mixed together and were energized by waves and lightning, volcanoes and sunlight, and they formed larger molecules. Molecules combining in more sophisticated ways were able to perform more sophisticated actions. Some molecules could form spheres; some could imprint their form onto other molecules; some could hold an electrical charge. The molecules must have come together in countless billions of unavailing ways and broken apart again before they came together in just the right order to form a working cell. The universe had learned something new: L.I.F.E.

"All living creatures are made of tiny units called cells," said Mrs. Rosetta. "Repeat after me: C.E.L.L."

"C.E.L.L.," said the cells.

It had taken a long time for chaotic molecules to form the first cell, but the first cell quickly began imprinting its order onto the chaotic molecules around it. It said to them: "Repeat after me: C.E.L.L." And the molecules said: "C.E.L.L. C.E.L.L. C.E.L.L. C.E.L.L." Before long the whole planet was covered with cells.

"Another important word of biology is evolution. Scientists think that life started out as something very small and simple a long time ago. They think that life has slowly changed into bigger forms. Repeat after me: E.V.O.L.U.T.I.O.N."

"E.V.O.L.U.T.I.O.N." Several of the children mixed up the last letters.

At first cells were very simple, but as they reproduced they occasionally mixed up the sequence of their genes and ended up with new forms and abilities.

"That's E.V.O.L.U.T.I.O.N.," said Mrs. Rosetta.

"E.V.O.L.U.T.I.O.N.," said the children.

Life slowly grew in size and complexity. Life learned how to put molecules together in the right way to capture and utilize the energy of the sun— P.H.O.T.O.S.Y.N.T.H.E.S.I.S. Cells learned how to put themselves together to form M.U.L.T.I.C.E.L.L.U.L.A.R. O.R.G.A.N.I.S.M.S. Organisms learned how to cooperate in societies. Creatures developed several ways of living, such as being P.L.A.N.T.S. or A.N.I.M.A.L.S. They shaped their bodies in various ways to enable them to move about in the water, underground, on land, or in the air. They developed senses for perceiving what was going on around them, and to process this information they developed a—

"B.R.A.I.N.," said Mrs. Rosetta.

"B.R.A.I.N.," said the brains.

Life learned how to put together sounds or molecules in complex ways so that creatures could share information with one another. At first such messages were simple and stereotyped, but as brains grew better at processing information they could create and uncode much richer messages. When humans evolved, communication became richer yet. Humans not only developed sophisticated spoken language, they invented something entirely new—written language. Starting with a few simple lines and pictures, humans developed thousands of pictures, and developed rules for combining those pictures in various ways to have various meanings. Then humans discovered an easier system, the A.L.P.H.A.B.E.T., in which a small number of abstract symbols represent the basic sounds the human voice can make. By arranging this handful of symbols variously, humans could represent everything in the universe, even as they discovered that the universe contained millions of different kinds of things.

"Who can define the word 'discover'?" asked Mrs. Rosetta.

Hypatia raised her hand. "Discover means to learn things, to find out things."

"Right," said Mrs. Rosetta. "Science has discovered the secrets of the universe."

Yet the alphabet was not entirely a new invention. Humans simply had discovered a new application for an ancient principle. This principle had been at work in the universe from the very beginning. It's the principle that combining a handful of elements in various ways can produce a dazzling diversity of forms and powers. All of cosmic evolution consisted of the working out of this principle on one level after another.

From quarks to atoms to stars, from organic molecules to life to DNA to brains, matter slowly, gropingly learned how to put itself together in new ways. The secret architecture of matter made possible a vast hierarchy of order, with one kind of order leading to another. If the nature of matter had been different, leading to too little or too much order, the universe might have gotten stuck in chaos or rigidity at any step of the way. The universe might consist of nothing but quarks unable to combine, or quarks glued into one giant block. Yet matter made possible just the right order and flexibility for quarks to build all the way to life. Sometimes the order of one level not only made possible the emergence of the next level of order, but forced it. More often, matter had to spend a lot of time and energy searching for the way to build the next layer of order. Each layer had its own rules, some quite complicated. Yet bit by bit, over billions of years, matter figured out its grammar and put itself together in eloquent ways. The blackboard of space filled with one object after another. And then the universe learned how to spell H.U.M.A.N.

"That's us," said Mrs. Rosetta. "That's the name of our species." Repeat after me: H.U.M.A.N."

As the humans spelled H.U.M.A.N., activity swarmed through the intricate system inside their heads. Electrical impulses raced down billions of neurons, fizzling out on some pathways, but on others adding up like tributaries flowing into a river. Messenger chemicals rushed back and forth between neurons. Inside the neurons, molecules swarmed, energy flowed, DNA spewed out instructions, and molecules broke apart and reformed with new meanings. Different sectors of the brain volleyed information back and forth.

The children spell 'human' with only five letters, but the universe spells human with an unimaginable number of letters, letters arranged into the intricate order that is the human brain. This order incorporates all the levels of order that have emerged in cosmic evolution. The quarks that bound themselves together in the first moments of the universe are in there. So are the atomic nuclei formed inside stars, and the molecules and cells that life has evolved over four billion years. Out of these simpler layers of order has arisen the ultimate order, a consciousness that can discern all the layers of order in the universe and the fourteen-billion-year process through which they all arose.

Atop fourteen billion years of order, the human brain began generating a strange new layer of order. Humans translated the language of the universe into human language. In the human brain the entire universe was constructed anew, but with something added. In the human brain the universe finally learned how to spell "I am." The universe cast a magic spell: "I am."

"And that's it," said Mrs. Rosetta—her forming of these words a continuation of the ordering process that began fourteen billion years before. In the head of each child, the magic loom swept back and forth, weaving her sound waves into meaning. "That's our spelling lesson for today."

In the Sky with Diamonds

After twelve hours of dyeing the sunlight blue, the air was thickening the blue into bright red, orange, yellow, and violet, colors that flowed steadily in tone and shape, glowing from clouds that also steadily changed shape. Within this sunset there was a smaller sky, the crystal sky within a diamond, and here too the sunset was occurring. Inside this carbon atmosphere the sunlight flowed through varied angles and densities which further scrambled its colors, giving this sunset a carbon-accented sparkle.

The sunset was also sparkling in the eyes of the woman who carried this diamond on her finger. She was standing on the top of a mountain and gazing over lower mountains and over hills that rolled downward into desert plains. She was watching the colors and the clouds flowing through the sky, and the patterns of light and shadow flowing across the land. Into her eyes the air-twisted light flowed and was curved again by her lenses, was focused to ignite within her brain a third version of the sunset, this one colored in both light and delight.

As the sunset began to fade, the woman turned and walked into the observatory. Her footsteps echoed from the huge dome. She walked through the dimness, past a dark and silent computer, past the telescope, and sat down at a control panel. She turned it on, and the panel bloomed with colored lights. When she reached for one of those lights and pressed it, the small amount of force in a human body, in the engines of a human finger, connected with much vaster and more powerful forces. To the throbbing of an engine, the eyelid of the observatory dome slowly opened to the sky. Another engine started pouring out the tons of power to turn the dome itself, which rotated past the sunset and almost stopped at one point of the sky. Then the telescope rotated to the same point and almost stopped—almost stopped but for a tiny continuing motion that would keep the telescope and the dome aligned to that point all night long as the stars flowed across the sky. In synchronizing the telescope to the turning Earth, whose circling was geared to the larger circles of the solar system, which in turn were geared to the larger whorls of the galaxy, the woman was touching her finger to the galaxy's luminous fingerprint.

Through the diamond on her finger she was also touching the larger forces at work inside the earth. The diamond contained the pressures that built mountain ranges and that triggered earthquakes and that melted rock and squeezed it upward through the crust and fired it from volcanoes. Yet sometimes magma was blocked from reaching the surface and it was pressured harder and harder, making it hot enough to break its molecular structure and bond it into new form. The diamond was pure carbon sieved from flowing rock and pressured until atoms locked into a tight recurring pattern, with each atom bonded to its four nearest neighbors in a tetrahedral arrangement. Geometrically this pattern was quite simple, yet like the geometrically simple markings on a cuneiform tablet this pattern contained a meaning far more complex and profound. In the untranslatable hieroglyphics of this crystal was written the ability of the universe to order itself.

As the colors of the sunset faded out, the stars faded in. A few brightest stars emerged first, then other stars, and then with each minute a dozen stars were separating from the twilight, then two dozen, and finally the sky was black and filled with thousands of stars, stars gleaming sharply in the clear desert mountain air, sparking white, red, yellow, and blue—stars sparkling like diamonds.

These diamonds too had crystallized out of formlessness, out of crude, flowing matter, out of fierce pressures and temperatures. Each star had been squeezed from a cloud of gas that had flowed through space and become trapped within its own gravity and collapsed inward, becoming denser and hotter until its atoms were colliding fiercely enough to break their atomic structure and fuse into heavier atoms. The bursts of light from each fusion merged into a tide of light flowing outward, a blade of light that cut away the rough, splotchy cloud and left a geometrically perfect star gleaming into space.

The diamond on the woman's finger was a continuation of the same formative process that on a larger scale had formed the stars gleaming above the observatory. This process had been continuing for billions of years, ever since it was unleashed by the Big Bang. It was toward the earliest traces of that process that the woman was aiming the telescope. Past the planets twinkling with today, past the stars and nebulae of our galaxy, past the galaxies glowing from the eons, she was aiming at one of the first galaxies that crystallized from the chaos of the Big Bang.

From the first moments of the Big Bang, the universe was rich with forces able to generate order. Some of these forces went to work instantly, and some had to wait for ages, wait for the Big Bang to sort itself out, before they could take effect. Elementary forces combined to form more sophisticated ordering forces. Some forces gathered matter together, some forces locked it together, some forces gave it shapes, and some forces, like gravity, could do it all—gather, hold, and shape matter.

In the Big Bang matter swarmed and collided intensely enough to trigger the strong nuclear force to bind particles into atomic nuclei. As the Big Bang expanded, the electromagnetic force bound electrons into orbit around the nuclei, forming atoms. Gravity was also binding the swarm into larger shapes, into galaxies, with quasar hearts and star veins winding throughout. Inside stars atoms were smashed together hard enough to bind them into heavier elements, including carbon. These elements slowly accumulated and were scattered into space by dying stars, forming clouds that then crystallized into new stars and into planets. Gravity, flowing out from a planet, and sunlight, flowing in, merged their energies to work upon the planet's surface; matter was sorted and mixed and piled and pressured into new forms, into rocks and mountains, raindrops and oceans, snowflakes and glaciers. Thousands of large and small processes brought matter together in new ways and let the electromagnetic force bind atoms into new substances, into simple molecules repeated over and over or into intricate collages of dozens of kinds of molecules.

The atom most capable of forming complex molecules was carbon. Each carbon atom could bind to four other atoms at once, and bind firmly enough to maintain a complex form, yet not so firmly that a huge amount of energy was necessary to break them apart. Brewed by ocean waves and volcanoes and lightning, carbon atoms grew into long, intricate molecules whose forms sometimes gave them a special function in the society of carbon molecules, such as being able to bind molecules together or split them apart, being able to carry a large amount of energy or information, or being able to form a wall or a sphere. After ages of random combinations, the right molecules came together and merged their tiny abilities into that ability called life.

The universe's most talented formative force was finally operating. A billion years might go into making a planet's first cell, yet that cell soon filled the whole planet with cells. Each cell was far more complex than any crystal, containing dozens of major forms, each composed of form within form within form. The order of life consisted not just of complex forms but of complex activities, activities that had to mesh perfectly in location and timing, activities of which a cell's forms were but momentary manifestations.

As mutations produced new genes from which cells could crystallize, the forms and activities of life became more and more sophisticated. The elementary binding forces that held atoms together were thickened into forces that could hold cells together, allowing cells to build bodies of dozens or trillions of cells, bodies with special shapes and skills. The formless, ignorant dust of a planet arose into a multitude of shapes and skills, into plants with a billion shapes for catching and binding the sunlight; into animals shaped for moving through water, on the ground, underground, or in the air; into bodies that could survive great heat or cold; into bodies to fit the shapes of mountains or deserts or jungles. These bodies arose not just once in evolution but arose again and again as each generation summoned the dust back into the forms of life.

This proliferation of forms was possible because the forces that give shape to matter and hold it together continued evolving. The force that binds cells together evolved into roots strong enough to bind a redwood tree to a mountainside, and branches reaching high and holding onto their leaves even through a tug-of-war with a hurricane; it became the bones supporting the flesh of an elephant, and the muscles that enabled a bird to flap its wings. The ability of a cell to draw in atoms and shape them into long molecules evolved into roots sucking in pounds of water and soil to be shaped into a tree; it evolved into animals with limbs and mouths for gathering food, and with a network of organs for processing food into the animal's shape.

Life's powers for binding itself together then evolved from working within organisms to working between them. Life evolved societies in which individuals, like molecules in a cell, performed different roles to help maintain the whole. Such cooperation required a new kind of binding force, a force not physical as the animals were not physically connected, but a force that could hold them together as strongly as the atoms in a diamond are bound. This strange, difficult new force was feelings. As feelings pulsed with different flavors, they bonded animals into pairs and families and packs and societies and nations. Because the continuation of life depended most upon bringing male and female together, between them this bond worked most strongly.

After glancing at her watch often through the evening, the woman rose from the console. She walked beneath the telescope, the brain-dreamed sunflower tracking its nourishment. She passed the computer, which now hummed with activity.

Inside that computer a galaxy was whirling, rolling its clouds and billions of stars through the silicon veins. That galaxy was emerging from the chaos of the Big Bang, flowing toward order, building lattices made of gravity and stars. From such lattices had eventually emerged the silicon lattices of this computer, silicon whose crystalline structure was the same as the structure of carbon in a diamond, with each atom bonded to its four nearest neighbors in a tetrahedral arrangement. The computer held maze within maze of crystals. The computer was a galaxy's offspring silicon galaxy, and it was carrying forward on a higher level that galaxy's quest for order; it was searching within the whirling galaxy for patterns, reading the galactic fingerprint to trace the passage of cosmic evolution, revealing to the humans who made it their own beginnings, the first steps of the process that formed the lattices of their brains, their thoughts, their feelings.

She reached the door, stepped outside, and stood gazing out at the stars and the vague shapes protruding from the darkness below. From that darkness she was looking for a light to emerge, winding its way up the mountainside.

As she waited she remembered another night when she had stood here and gazed down the mountainside, a winter's night when a sudden blizzard had closed the sky and the dome. The wind had chilled her face and run its icy fingers through her hair. The snow had fallen heavily from the sky and from the universe's genius at creating form, but that night there was so much form it became a chaos. She had gazed anxiously into the blizzard, and after too long she saw a light flicker from it and slowly grow into two lights like eyes steadily approaching, as if emerging from the same primordial forces shaping the snowflakes. As the car had pulled up to the observatory she saw the snowflakes pouring past the headlights, twisting the light into sparkles.

But tonight a clear darkness divulged the car. As the car drove into the parking lot, the woman headed down the steps and toward it. A form emerged from the car, and amid thousands of stars and the gravity bonding and guiding them, the two humans met and embraced and kissed, and each verified with three words that the emotional gravity holding them together and guiding them on course was still working.

Holding hands, they walked up the steps and into the observatory. They passed the telescope and its photographic bull's-eye into which the star-fired arrows were flowing. That photographic plate had started out as empty and black as the universe once had been, but with the passing minutes of a long exposure, order was emerging from the blackness. The sharp white dots of foreground stars emerged first, then the vaguer white of a galaxy.

Within the woman too, a form was slowly emerging. Her womb was a photographic plate whose chemicals had been activated to form the image of herself. Through a long exposure lasting nine months this plate would collect bits of information, collect them from out of a past as vast as the eons from which quasars shine, collect them from the genes containing the whole evolution of life on Earth. As the shapes within those genes shined through into the womb, there was appearing the shape of a human, her head and legs and arms curled up while she was waiting to be born, curled up like the spiral arms of a galaxy, like the billions of galaxies that had lain curled up in the womb of space waiting to be born into life. She carried a starchild.

The light flowing into the telescope above the woman was also flowing into the lens upon her finger. Through the opening in the observatory dome the starlight poured, light from a hundred stars visible to her and from countless stars her eyes couldn't register. It came from nebulae glowing with the light of newborn stars; it came from stars billions of years old; and it came from stars that had died into neutron stars and white dwarfs. It came from the whole Milky Way galaxy spread across the sky. It came from a million other galaxies, from out of journeys of billions of years. It came from galaxies still forming from the Big Bang. Light from the entire history of the universe was flowing into her diamond, being folded this way and that and emerging again in the faintest sparkles.

Her diamond contained the whole universe, not just in light but also in its own structure. Its atomic nuclei contained the energy of the Big Bang. Its carbon atoms contained the fire of the stars. Its crystalline form contained the geological fires of Earth. Its cut and polished form contained the evolution of life into complex thoughts and feelings, into a form of carbon that could choose as its symbol of continuation a form of carbon both strong and beautiful. The diamond was the pure element of life, the element that had made the universe's most amazing adventure into order. The diamond embodied the whole mystery of form in the universe. In the starlight flashing from it, as in the starlight entering the crystal eyes of humans and being folded into consciousness, that mystery was proclaiming itself as clearly as it had in the flash of the Big Bang.

The Hubble in the Bubble

Once again I arrived in the birthplace of astronomer Edwin Hubble, the discoverer of the Big Bang and the expanding universe. This visit offered some extra entertainment, for tonight the Marshfield high school was putting on a musical based on the works of Dr. Seuss.

Hubble lived in a bubble
But he wouldn't know it for years to come;
Hubble was just a boy
Throwing baseballs and chewing gum.
Hubble blew a bubble,
Blew it so big he was turning blue,
And from this double bubble
Hubble thought he heard a Who.
A tiny voice was calling out:
What's this universe all about?

Hubble was in a bubble, the bubble of his mother's womb. Hubble started out as a singularity, a dot that contained fantastic powers for order, a biological universe that expanded steadily. Out of formlessness arose spiral upon spiral, cell upon cell, shape upon shape, system upon system, ability upon ability—out of formlessness arose a head and a face. Out of blindness arose eyes; out of silence arose the ability to hear and to speak; and out of mindlessness arose a brain that would sift through its synaptic order the order of the universe and find in it an order that no human had ever found before, a brain that would be a key for opening the secrets of the universe.

Yet if the universe had been more orderly, Edwin Hubble never would have been born. If not for a serious outbreak of chaos, Edwin's parents never would have met. The Hubble family owned large orchards in Marshfield, Missouri, and their main crop was apples, the fruit that had ruined the perfect order of the universe and brought chaos and death into human life. It was the small chaos of an apple falling onto Isaac Newton's head that prompted him to discover the gravitational order of the universe. Yet for John Hubble gravity meant only chaos when the team of horses he was working in the fields bolted and knocked him down and dragged him—ensnarled in the harness—across the rough and rocky ground, grinding his hands and face and chest, shredding his overalls, leaving him laying there a bloody pulp, barely able to see the sky through his swollen, muddy, bloody eyes. He was taken to Marshfield's

41

Dr. James, who sent for his daughter Virginia to help, but as soon as Virginia walked into the room and saw the bloody mess she felt ill and threw up. She recalled later that she "never wanted to see John Hubble again." But for John Hubble, Virginia signified not chaos but kindness and help. He fell in love with kindness and with her and wooed and married her. She threw up again over the expanding universe of Edwin Hubble within her.

Dr. James encouraged Edwin's interest in astronomy by building a telescope for his eighth birthday. Edwin declared that instead of the usual party full of candles he just wanted to stay up way past his bedtime for the birthday candles in the sky. He stayed up for hours steering the telescope from planet to star to nebula, fascinated to discover the depths and the details behind the sky the eye could see. Edwin was especially fascinated by Mars, for this was the time when Percival Lowell was insisting that Mars held a vast system of canals built by an ancient and superior civilization. Two years later Edwin stayed up all night to watch a total eclipse of the moon.

Edwin grew up in a good time and place in which to become interested in the sky. It was before the invention of movies or television, and the moving lights in the sky were still the most dramatic light show around. Edwin lived far from a city with its lights and entertainments and distractions. Eighty miles away from Marshfield in another small Missouri town another boy, Harlow Shapley, was also noticing the sky, and the two of them would grow up to become the two most important astronomers of the first half of the twentieth century.

Hubble and Shapley came along at a good time in which to make discoveries about the universe, for the industrial age was placing into human hands major new scientific tools. Since Galileo's time telescopes had improved only slowly, but now telescopes were making large jumps in size and quality. Photography was enabling astronomers to far transcend the human eye. Instruments like spectroscopes were revealing the inner physics and the largest motions of the stars. Advances in physics and chemistry were offering surprising connections with astronomy. Edwin Hubble found the most perfect time and place for astronomy when his father, for the sake of his insurance business, moved the family to Chicago. Edwin enrolled at the University of Chicago, which recently had built Yerkes Observatory with its world's largest telescope; Edwin started assisting the Yerkes staff. But Edwin's father was strongly opposed to him pursuing an impractical career in astronomy and wanted Edwin to go into the law, the laws of humans and not the laws of stars, the laws that no longer worked as well as the laws that ruled the sky, the laws that couldn't stop humans from committing theft and murder. Edwin went off to Oxford to study law, and there he adopted an English accent and aristocratic airs, sporting a black cape and a cane and a pipe. But aristocratic airs included defending a gentleman's honor, and Edwin got himself into a pistol duel when a man decided that Edwin was too attentive to the man's wife. Both men missed, saving Edwin from ending as a mere tiny and ignominious illustration of the laws of trajectory. It was only the unexpected trajectory of his father's death that enabled Edwin to abandon law for astronomy.

Hubble was offered a coveted job at Mt. Wilson Observatory, but he was more eager to join the army and see World War One, and he skimped on his Ph.D. dissertation so much that he had to redo it after the war. Hubble rose to the rank of major and always claimed that he had commanded troops in frontline trench combat, but Hubble's biographer Gale Christianson could find no evidence for this in Hubble's military record, and in fact Christianson found a letter in which Hubble complained that he'd seen very little action. But at Mt. Wilson Hubble went by the name of "Major" and wore his army boots and trench coat to the telescope.

Hubble did make Major discoveries. He proved that nebulae—long thought to be only local gas clouds—were actually other galaxies far outside our own, and that all the galaxies were racing away from a common origin. The universe actually had a birth, and was growing up.

Hubble lived in a bubble
Blowing larger very fast;
Hubble stood on the rubble
Of the astronomical past.
Long gone was the solid Earth
At the center of cute little stars.
Now Earth was but a speck
Flying with galaxies, moons, and Mars.
Now humans were made of stuff
That started off with a Bang
And was cooked and changed enough
In stars and volcanoes and cells
That finally it looked at the night and knew
That humans were a Who.

Since Copernicus the human conception of the universe had been changing slowly, about as slowly as telescopes were growing, but now within a few decades, a dizzyingly short time in cultural terms, the universe was transformed. The universe became vastly larger than humans ever had imagined. The universe became even more ancient than in the most poetic imaginations of Hindu sages trying to stun humans into humility. The universe became incredibly dynamic, filled with motions and transformations. Matter became loaded with energy and talent, crafting itself into form after form, system after system, including all the forms and systems of life. Instead of human life being its own little story and the universe being "out there" as a pretty decoration or a foreign country, human life was thoroughly embedded in a dramatic cosmic story. A cell was just as much an expression of the cosmos as was a galaxy. A human brain was another kind of star burning with exotic energy. To conduct a thorough inventory of the universe Edwin Hubble needed to turn the telescope upon—Edwin Hubble. The most telling observations of the nature of the universe might come not from scanning the Mars fields, but from looking at Marshfield.

For a long time after Edwin Hubble transformed the universe, travelers passing through Marshfield saw no visible acknowledgement of him. Indeed, his entire home state seemed to have forgotten him. In the 1980s I offered to write an article about Hubble for Missouri's glossy, state-themed magazine. The editor had never heard of Hubble. I explained that Hubble had discovered the expansion of the universe. The editor didn't recognize this as a significant discovery. He suggested that I come up with a quote from a famous astronomer saying that this Hubble guy had done something noteworthy. But, I explained, Hubble was the most famous astronomer of the twentieth century. The editor replied: says who? I gave up, and the magazine went on doing articles on Missouri-born movie stars. This was before the Hubble Space Telescope was launched and made the public aware that someone named Hubble must have done something noteworthy. Only in 2002 did the state of Missouri add a bust of Edwin Hubble to its "Hall of Famous Missourians" in the state capitol.

A few years after the Hubble Space Telescope was launched, signs showed up on the interstate highway through Marshfield announcing that this was the "birthplace of astronomer Edwin Hubble." This stretch of the interstate was now "The Edwin Hubble Memorial Highway." Billboards invited travelers to stop and visit a replica of the Hubble Space Telescope on the town square. Marshfield put quite a bit of effort and money into building its telescope, which was one-fourth-scale, twelve feet long, and 1,200 pounds of stainless steel (but no optics inside), perched on thin stilts to give it the feel of floating in space. The telescope opening was sealed with glass to prevent it from becoming a giant pigeon nest with very earthy streaks and smells. Marshfield built the telescope more as a tourist attraction than as a fond tribute to a favorite son. The truth was that Edwin Hubble had been largely forgotten in his hometown, except perhaps for one old man who still recalled young Edwin staying out at night with this telescope.

The house where Edwin was born was torn down around 1970. The man who demolished the house saved the antique doorknobs and later donated them to the town museum. Then he heard about some Hubble relic being sold on e-bay, and he took the doorknobs back. The town museum couldn't afford to bid for the doorknobs, so the only Hubble artifact in town left town.

It turned out that the replica Hubble telescope wasn't very successful at luring tourists into town. A decade after the replica was built, many of the storefronts around the town square were empty. Marshfield thought the telescope was a winning idea because Marshfield was on old Rt. 66 and Rt. 66 nostalgia and tourism were thriving, and the space telescope seemed a worthy match for the old, quirky, homespun tourist attractions along Rt. 66. But it seemed that Rt. 66-minded tourists were more interested in rundown greasy-spoon diners with faded pictures of Elvis on the wall than in high-tech astronomy. After all, Rt. 66 nostalgia was less about greasy-spoon diners than about frustration with an impersonal franchised world where every meal tasted the same and no one knew your name; it was a longing for a mom and pop who designed their diner architecture and menus with imagination and uniqueness and who dearly cared about their customers. It wasn't clear how this longing for a personal and caring universe would be satisfied by a telescope and an astronomer that had revealed a universe that was far less personal than a McDonald's drive-thru. The tourist stops by the interstate sold plenty of Rt. 66 souvenirs, but they didn't offer any Hubble souvenirs.

Marshfield's town square was typical of hundreds of American town squares, with a court house, various markers and monuments, and a World War Two cannon. And then there was the Hubble Space Telescope, floating there and pointing slightly skyward. The startling anomaly of it suddenly transported the town square out of the realm of Norman Rockwell Americana and into a cosmos of rushing majesty and mystery, a cosmos whose columns were not just concrete but glowing nebulas, whose laws were written not just in books but in atoms, and whose judgments of fate were pounded out not just with a gavel but with comets and supernovas.

I first saw the town square in autumn: its tall trees were still plump with leaves but they were starting to shift toward the autumn end of the spectrum, the redshift; soon the trees would be flying thousands of red flags of the expanding universe, until entropy forced them to surrender. For now the leaves were mostly green and still pointing at the sun, soaking up energy just as the cells of Earth had been soaking up the sunlight for billions of years, transforming that energy into structure and action. Now Earth's cells and leaves had evolved into the solar panels of the Hubble Space Telescope, which now transformed the energy of the sun into sophisticated imaging of the universe, and into the images pulsing within the human brain.

Beneath the telescope was a plaque, which explained that Edwin Hubble was born in Marshfield in 1889, but when it tried to explain the Big Bang universe it got a bit blurry. Just like the vision of the real Hubble Space Telescope.

The sign also noted that the replica had been dedicated on July 4, 1994. This was over fours years after the Hubble Space Telescope had been launched. It was about six months after NASA launched a space shuttle mission to repair the defective Hubble, a mission that captivated the public imagination and restored some of the sense of heroism originally connected with the Hubble Space Telescope. To drum up enthusiasm and $2 billion to pay for Hubble, NASA had raised public expectations for it very high, only for Hubble to come crashing down when it turned out that its mirror had been polished to an incorrect curvature, all because a simple calibration device had been used incorrectly. NASA hadn't bothered to test the telescope before launching it. Hubble's vision was seriously blurred. The scientific community was shocked and incredulous; TV comedians told Hubble jokes; and the public regarded Hubble with scorn. In Marshfield there was no rush to associate the town with the telescope. Only six months after a repaired Hubble started sending magnificent photos into daily newspapers did the Marshfield replica appear, and with 4th of July pride. Other signs of pride included an elementary school changing its name from the bland "R-4" to "Hubble," and the town newspaper placing on its masthead an image of the Hubble Space Telescope and the tag "Home of Edwin P. Hubble." At least for awhile: later on the newspaper replaced Hubble with an American flag.

Do humans ever take hometown pride in the universe? Patriotic pride in real stars and comet stripes?

A few years after the telescope replica appeared, I stopped in Marshfield to visit a friend who was originally from there and who had now returned to live near family. Lloyd's engineering career had taken him to the Goddard Space Flight Center, which later would be in charge of building the Hubble Space Telescope. In the 1960s Lloyd had been instrumental in building the worldwide network of giant tracking antennas used for the manned space program. Television viewers of the Gemini and Apollo flights became familiar with these tracking stations as capsule communication was passed from station to station, from Spain to Australia to America. It was through Lloyd's antennas that one small step on the moon was witnessed by all humankind.

From cave paintings to Rt. 66 Burma Shave jingles to moon whispers, humans are geniuses at communicating. We have even learned to speak the radiant language of galaxies and black holes. We have even built artificial voices so that a Stephen Hawking can speak from the black hole of his own body.

But sometimes human cells go blurry and stop communicating even with their neighbor cells. Both Lloyd's wife and my father had disappeared into the black hole of Alzheimer's disease. For two years Lloyd and I had helped out at the same nursing home dining room table. Lloyd's wife was often as distant and silent as the moon, and he had to coax her to every bite. All the technological skills that had allowed communicating across a quarter million miles of space couldn't penetrate the imperfections of our own biological substance. My father too had once been a master of communication, working for Bell Labs in their transistor-inventing heyday and writing several textbooks on cybernetics. But now the logistics of eating a sandwich would defeat him, and I had to coach him through it.

Sometimes I brought a magazine to the table. The photos offered a conversation piece while we waited for meals. One time I showed my father an astronomy magazine with the latest photos from the Hubble Space Telescope. I think he enjoyed the colors, but otherwise he didn't grasp the meaning of it. More than once I tried to escape the dining room sound and fury by reading about some new astrophysical theory, but sometimes I couldn't grasp its meaning much better. Surrounded by wheelchairs and incoherent limbs and voices, I couldn't help noticing that the bizarre rules of the universe weren't confined to the sky, that we all struggled against the gravity that pulled us into darkness and death.

Now both my father and Lloyd's wife were gone. Now Lloyd and I talked about their ordeal, about the kindnesses and imperfections of the nursing home. Once we talked about Edwin Hubble, who came home from the chaos of war and pursued a greater vision. We talked about building NASA facilities in Appalachian hollows whose hillbillies had barely heard about the Apollo program, where outsiders were presumed to be hunting for moonshine. And moonshine they found, moonshine and Earthshine mirrored on the visors of astronauts seeking a vision of human life greater than our daily confusion and pain.

Before leaving town I stopped at the telescope. I wished I could see it at night, when you could walk underneath it and see its shape floating amid the stars, tracking the constellations. By day the telescope mirrored a closer part of the cosmos. Its round stainless steel offered a circus-mirror depiction of people, but it was not a well-polished mirror and offered only a very blurry, stretched-out image of basic colors and shapes and actions. When I first looked into it I didn't even recognize myself, so I waved to test it out, and a blurry hand waved back at me. I smiled, and I saw a vague smile, like Stephen Hawking's grotesque yet still genuinely happy smile, an enigma more terrible than that of Mona Lisa's smile. And yet, out of the strange and flawed matter that we are, we have reached out, with both our eyes and our Apollos, and won some of the insight of gods.

Edwin Hubble's fame made him a star in the celebrity world of Los Angeles, which was near Mt. Wilson Observatory. One night Hubble and Aldous Huxley and Frieda Lawrence, the widow of D. H. Lawrence, drove merrily over to the house of Harpo Marx, where Huxley pondered writing an astronomy-themed screenplay for a Marx Brothers movie. Huxley imagined the Marx Brothers up at Mt. Wilson Observatory, pushing the wrong buttons and sending the machinery berserk. Then Groucho Marx would gaze into the big telescope Hubble had used to discover the expanding universe, and Groucho would see simply the grinning, mute face of Harpo.

On another visit to Marshfield I visited the county history museum, in an old Carnegie library with its typical homey fireplace. The museum displays were rather homey—random collections of old farm tools and Boy Scout stuff. One display case was devoted to Edwin Hubble and included commemorative postage stamps; patches from the space shuttle missions that had launched and repaired the telescope; magazine and newspaper articles about Hubble and the telescope; and books about Hubble family genealogy—actually the "Hubbell" family, Edwin's branch having altered the usual spelling.

The corridor of the nearby courthouse held an autographed poster of the space shuttle crew that had repaired the Hubble Space Telescope, and a 1917 newspaper article about how Edwin, visiting from Chicago, had walked twenty-two miles in four hours to see his grandfather.

This happened to be the weekend of Marshfield's annual autumn "Harvest Days" festival on the courthouse lawn. If Harvest Days started out as a farmers celebration of the harvest, then maybe the Hubble family brought their apples and apple pies to show off. Today the Harvest Days booths offered mostly country crafts, beauty products, and junk food. But something special was happening this year, something that ordinarily wouldn't happen in towns this small: NASA had sent a tent-full of exhibits, including a space suit, models of spacecraft, and a genuine moon rock. School buses from dozens of miles around were letting off flocks of kids.

I stared at the moon rock. As a boy Edwin Hubble had stared through his telescope at this very rock, at a moon that was symbolic of the unreachable and mysterious universe. Edwin received his eighth-birthday telescope six years before the Wright brothers flew. Edwin Hubble may have greatly expanded the universe, but technology had shrunk our local portion of it.

The moon rock was safely encased in a Plexiglas bubble atop a solid pillar, but the Marshfield sheriff was taking no chances: a deputy was sitting right next to the moon rock, and at night it was locked up in a bank vault across the street. I asked the deputy if he knew that Edwin Hubble's mother was a James, related to Jesse James, against whom no bank vault or deputy was safe, and he laughed.

The school kids stared at the moon rock and asked if it was really the moon. "Does it glow in the dark?" The moon rock stared at the Earthlings and asked if the universe was really alive.

School groups took turns gathering around the Hubble Space Telescope replica. One teacher told her group that Edwin Hubble was the designer of the Space Telescope. The next speaker emphasized that the telescope was merely named for Edwin Hubble, who didn't have anything to do with inventing it, but his audience seemed to have heard otherwise, for two kids asked him about Hubble designing or building the telescope. One kid eagerly asked about the World War Two cannon, and the teacher explained that this wasn't part of the NASA exhibits and we wouldn't be shooting it off today. No Big Bang today.

On the opposite side of the courthouse from the Hubble telescope I found a booth where a Native American was selling herbs. Steve Yellowhand was a Lakota from South Dakota, but he had lived in Marshfield for twelve years. I asked him how he had come to live so far from any Native American community, and he was a bit vague about it, but he admitted that he felt pretty isolated in Marshfield. Though herbs were his business, this was the first time he'd tried selling them in public in town. Steve was setting up a microphone to play his Native American flute, and he told me quizzically that the sheriff had told him to play quietly so he wouldn't disturb anyone in the courthouse, even though the opposite lawn was packed with yelling kids and loudspeaker talks about Edwin Hubble. When Steve had first moved to town the sheriff had tried to shut down his sweat lodge as a fire hazard, but then the fire marshal gave him a permit.

In his cosmology too, Steve Yellowhand lived in a separate universe. He told me that he was a dance chief for the Sun Dance, and he lifted his shirt to show me two long, serious scars on his chest. The Sun Dance was the most important Lakota ceremony, rich in symbolism of the Lakota cosmos, a cosmos in which the sun and moon and stars are supernatural beings who hold great power over human life.

At the astronomically right time of summer the Lakota build a ceremonial camp just for the Sun Dance. In the center they place a cottonwood pole, with a red banner on top symbolizing the sun. Many elements, such as buffalo skulls and a sweat lodge, are arranged in symbolic directions. The Sun Dance is partly a warrior ceremony in which young men prove their toughness, and partly a vision quest for seeking supernatural wisdom. The dancers form a circle, symbolic of the cosmos. They stare at the sun.

In the most extreme form of the Sun Dance the dancers pierce their chests with sticks, tie ropes between the sticks and the cottonwood pole, and strain against the pole, even flinging themselves into the air, until their chest muscles are ripped. The bloodiness of the Sun Dance is repugnant to many other Native American tribes. When Christian missionaries objected to the Sun Dance, the Lakota started claiming that it was actually a Christian ritual, a reenactment of Christ's crucifixion on a tree.

Steve told me that the Lakota word for god translated as "the great mystery." I asked him what he thought of Edwin Hubble's approach to the universe, and he answered that the Hubble Space Telescope was okay since god wanted us to look at his beautiful creation. But Hubble and his telescope only saw "the beyond," while the Lakota god was "beyond the beyond." Then Steve added: if it turned out there wasn't any god, it was still good for us to look at the cosmos.

On the visit when I saw Dr. Seuss in Marshfield I just missed the dedication of a sesquicentennial mural on the town square, right in front of the telescope replica. The main mural motif was a giant American flag, and in the blue corner with its white stars floated the Hubble Space Telescope, implying that the Hubble Telescope belonged not to scientific space or—heaven forbid—existential space, but to patriotic space. From the blue corner, red and white stripes flowed through the rest of the mural and were studded with depictions of local events, such as the 1880 tornado that leveled the town and killed sixty-five people, and presidential visits from Harry Truman in 1948 and George H. W. Bush in 1991. The mural included depictions of Edwin Hubble, a Rt. 66 highway sign, and a local Blue Jay basketball player. This was how one little cosmic spot conceived its identity some fourteen billion years after the Big Bang.

It was Saturday night and the big event in town was the Marshfield High School production of *Seussical the Musical*, a hit Broadway show based on the works of Dr. Seuss, who wasn't a real doctor but who had dropped out of Oxford, Edwin's Hubble's alma mater, frustrating his father's ambition for Seuss to be a "Dr."

As I waited in the auditorium lobby I heard an unnaturally loud voice. I watched a young boy—the same age Edwin had been when he lived here—patting strangers on the back and asking a very friendly, "What's your name?" Then the boy tried awkwardly to carry on a friendly conversation. It was obvious that the boy was mentally handicapped. In his case, it seemed that being mentally slow meant not recognizing that the universe isn't always such a friendly place. It turned out that the boy sat right behind me, and when the show started he continued making exclamations and comments, loud enough to be heard by much of the audience. The boy's mother tried repeatedly to quiet him, but he just went on expressing his amusement and puzzlement, oblivious to everyone else, and finally his exasperated mother gave up and took him away.

Seussical the Musical was a blending of several favorite Dr. Seuss stories, but it was mainly about Horton hearing a Who, hearing a whole world in a speck of dust on a clover leaf, people who are people no matter how small. The show was narrated by the impish Cat in the Hat, who offered his philosophy that this might be a crazy universe, but things could be worse.

Edwin Hubble owned a black cat named Nicholas Copernicus, although the cat seemed to think it owned him. Nicholas refused to play with toy mice but loved playing with Edwin's furry pipe cleaners. Once when astronomer Fred Hoyle was visiting and declaiming from Macbeth, his sudden exclamation caused Nicholas to leap straight up and race outside into the night. During World War Two Hubble used his scientific training at a ballistics testing base—the Big Bang turned into artillery shells soaring and falling and blowing up—and the Hubbles lived in an old house they dubbed the Haunted House, which was haunted at least by mice, so they got a half-wild cat to bring order, except that the cat soon had two litters and now nineteen cats were running about the house. Edwin got a copy of T. S. Eliot's *Old Possum's Book of Practical Cats* to read to them. When the Hubbles moved away Edwin concocted a story that these cats had a historic bloodline, and thus he was able to place them in the fanciest homes.

The Cat in the Hat sang whimsically about this kooky cast of characters; Horton declared that the universe contained Whos on specks of dust; people declared that Horton was crazy; but Horton insisted that even in an existential universe where people are a very tiny thing, they still held importance and dignity.

At intermission many of us went outdoors to watch a spectacular lightning storm in the distance. A major storm front was moving through; eerie heavy clouds were racing across the full moon; the wind was gusting; a tornado had touched down thirty miles from here and tornado warnings were out all around us. During the second half of the show I was half-prepared for sirens to sound and for a tornado to try to carry us off to Oz—except that we were already in Oz.

The Cat in the Hat sang whimsically about life in the expanding universe of Edwin Hubble. The kooky cast of characters sang and danced and paraded in whimsical costumes before whimsical buildings and trees. Oh, the things you can think, oh the verses and curses, when you think about expanding universes.

When the show was over, the parking lot was full of puddles, each mirroring eerie clouds racing across the full moon. Leaving a sturdy school building for a flimsy wooden mom-and-pop motel made me feel rather vulnerable. I sought out mom-and-pop motels because they represented a friendlier universe.

As I was approaching the town square I recalled my old wish to see the Hubble Space Telescope at night. This would be my only chance. The square was deserted. I got out and walked up to the telescope. It vaguely acknowledged the full moon. It probably answered the lightning too. I circled the telescope and then paused when the telescope became a large mysterious shape framed against the eerie clouds racing across the full moon. An eerie moon, eerie clouds, and an eerie telescope were blowing through space long after the ultimate lightning storm had set them loose. Now the darkness held eerie eyes, eerie smiles, eerie minds. Oh, the things you can think.

When I drove away into the eerie night the Hubble Space Telescope must have seen my tail lights, my redshift, as the universe expanded ever farther.

The Hubble House

Recipe for a Hubble Hamburger: Begin with a Big Bang. Cook at 10,000,000,000,000,000,000,000,000,000,000 degrees for a microsecond. Quickly reduce heat by expanding the universe. Allow physical forces such as gravity and electromagnetism to separate. Remove all antimatter through annihilation with matter, leaving residue of matter only. After one second reduce heat to 100,000,000,000,000,000,000 degrees. Allow protons and neutrons to form. Continue reducing heat until electrons combine with protons to form hydrogen atoms.

The waitress placed before me a glass of H2O, hydrogen and oxygen.

Here in Minnesota they take water for granted. They have so many lakes that the state's license-plate motto, "Land of 10,000 Lakes," leaves out far more lakes than we ever had in my own state of Arizona, the land of drought-stricken, bulldozer-made reservoirs. In some places in the Southwest it's illegal for a waitress to serve a glass of water unless the customer requests it. On Mercury and Mars you couldn't get served a glass of water if you were dying of thirst. I lived in a corner of the solar system where we appreciated water, so as I picked up my glass of water I paused to appreciate it. I looked at hydrogen that was formed in the Big Bang and that still sparkled with light. Silently, since I was alone, I offered a toast: "To the Big Bang, for creating hydrogen." I put the glass to my lips and felt the smoothness of hydrogen. My throat and my cells felt grateful for the Big Bang.

I was eating lunch at the Hubbell House restaurant in the small town of Mantorville, Minnesota. The Hubbell House was started in 1854 by the same family that gave birth to astronomer Edwin Hubble, the discoverer of the expanding universe.

In 1640 Richard Hubball came from England to New Haven, Connecticut, and from there the Hubbell family spread out across America. They also spread out in how they spelled their name. They left their name on American history in diverse ways. New York Giants pitcher Carl Hubbell got into the Baseball Hall of Fame for calculating the acceleration and gravitational deflection of spheres. John Lorenzo Hubbell's Arizona trading post was turned into a national monument. In the age of Thomas Edison, Harvey Hubbell invented the electrical plug and wall socket and the pull-chain light switch that let there be light. William Wheeler Hubbell invented the artillery-shell fuse used by the U. S. Army in the Civil War and for decades to come—used for lots of big bangs. The Hubbell name reached its highest orbit with the

Hubble Space Telescope, named to honor Edwin Hubble. When the Hubbell Family Historical Society designed a logo it consisted of Richard Hubball's sailing ship and the Hubble Space Telescope. You can read all about Hubbell family history in a book on a shelf in "The Hub," the wine room in the sprawling, many-roomed Hubbell House restaurant. When the Hubbell Family Historical Society held its first national family reunion in 1981, they met at the Hubbell House.

The Hubbell House was founded by J. B. Hubbell, who, in spite of the best efforts of Hubbell family historians, remains a mystery man. He mysteriously showed up in the new village of Mantorville in 1854. The Hubbell House brochure says that J. B. Hubbell arrived on a stagecoach; the Hubbell family history book in "The Hub" says that he arrived on a wagon train. Neither the Hubbell House brochure nor the Hubbell book repeats a report that J. B. Hubbell was a thief running from the law in Illinois. In Mantorville J. B. Hubbell became the first county sheriff; legend says he also ran an illegal liquor operation from the Hubbell House. For his first hotel and stage stop, J. B. built a 16x24 foot, two-story log building. Two years later he built the three-story stone building still used today. It's a mystery where he got the money for this, but legend says that unpayable debts drove Hubbell to leave town a few years later— no one is sure which year. Then J. B. Hubbell disappeared. Legend says he headed west, went grizzly bear hunting, and was never seen again.

J. B. Hubbell's restaurant kept his name and kept going without him. It benefited from being on a major stage route to the Twin Cities. In 1876 President Ulysses S. Grant stopped here, signing his name on a guest register now proudly framed beside the front door. Today the Hubbell House uses its Civil War-era history, its walls loaded with historic photos and artifacts, to attract gourmet diners from nearby Rochester and from the Twin Cities an hour's drive away. The hotel is long closed, but it is still haunted by a ghost, supposedly that of a stage coach passenger.

When I looked at the menu I noticed that several items included the name "Hubbell." Since I wanted to get into the Hubbell spirit I decided I would choose from these. For lunch the choice was between the Hubbell Hamburger and the Hubbell Club Sandwich. For dinner I could choose between the Hubbell Chicken Kiev, Hubbell Shrimp Fettuccine, Hubbell Chef Salad, or Hubbell Prime Rib. For a side dish I could have Hubbell Wild Rice. For dessert the only choice was the Hubbell Cheesecake.

Recipe for a Hubble Hamburger (or Hubbell Prime Rib, Hubbell Wild Rice, Hubbell Cheesecake, etc.—the recipe is the same for all items, except for final steps): Cook inside stars for a long time. Minimum temperature: 10 million degrees. Collide protons with sufficient energy to fuse into atomic elements. Increase heat to 200 million degrees to create carbon. Don't rush cooking of carbon, the most essential ingredient. May require billions of stars for billions of years. Continue increasing heat until iron is produced—essential for making stoves and stainless-steel utensils. Use supernovas to cook elements heavier than iron. Spread all elements into space and swirl in clouds.

The waitress set my Hubbell Club Sandwich before me. It sure was thick and stuffed with stuff. The bread was marbled, with intricate brown and white swirls. The swirls reminded me of interstellar gas clouds. I looked at the glass of hydrogen next to the sandwich and saw the vast distance between them. The hydrogen was the first and simplest element of the universe, created in the Big Bang. The sandwich was stuffed with complexity, the product of billions of years of cosmic evolution. I picked up the sandwich and opened my mouth and tasted fourteen billion years. It was good.

I had come to the Hubbell House to check out a curious fact: Edwin Hubble's assistant astronomer, Milton Humason, was born only a few miles away from the Hubbell House, in the town of Dodge Center, Minnesota, in 1891. When Milton Humason was a boy perhaps he came to the Hubbell House and ate lunch.

It was Milton Humason's hard labor that enabled Edwin Hubble to discover the expanding universe. Starting around 1920 Hubble and Humason worked together at the new Mt. Wilson Observatory in southern California, which had the world's most powerful telescope. Humason took the galactic photographs and spectrograms that Hubble studied to determine the distance scale of the universe; Hubble found that the galaxies were redshifted, spreading out from a common origin, the Big Bang. Even through the Mt. Wilson telescope most galaxies were too faint to be seen by the human eye, and Humason had to expose a photographic plate all night long—even all week long—to obtain a usable image. Great skill was required for such photography, but also sheer endurance. Humason had to keep the telescope moving steadily all night long, keeping a guide star in the telescope's cross-hairs. The motion of the eyepiece sometimes required an astronomer to become an acrobat, contorting head and body into awkward positions. Since the observatory dome was open to the night air, winter astronomy could be brutally cold, even in southern California. The physical ordeal of astronomy did not always correlate well with the dignity of astronomers, many of whom came from upper-class families and then obtained PhDs and professorships from the world's most elite colleges. Edwin Hubble came from an ordinary Missouri family but his Oxford education left him even more pretentious than most English Lords. Perhaps what Edwin needed was a good working-class bloke to shoulder the burden for him. Milton Humason had been working as the observatory janitor for $80 a month.

Even as a janitor Milton Humason earned a reputation for being meticulous. He had to keep clean the high-tech astronomy equipment as well as the ordinary floors. When Milton swept dust off the lenses and floors he couldn't perceive that this was actually stardust, elements that had been cooked up inside ancient stars. His future work at the telescope would help transform mere dust into stardust and give new dimensions to everything else about human life. He would help blow the smug roof off of every house on Earth and give them a Mt. Wilson view of an expanding and evolving universe, turning them into Hubble houses.

At least as a janitor Milton had been free to take a lunch break. As a galaxy photographer he had to stay at his post all night. Lunch might consist of a thick sandwich he kept wrapped in his pocket. Perhaps a thick club sandwich on marbled bread.

After lunch I headed up the hill to the Dodge County Historical Society Museum. It was housed in a former Episcopal church built in 1869, with the first pastor's wife still buried under the floor. I had already contacted the historical society and they had told me that they'd never heard of Milton Humason. He wasn't mentioned in any county history book. Now I asked the director if they might have any sources showing local Humason family history. She dug through their obituary records and came up with a batch of Humason obituaries, the most recent of which was 1918. The Humason family was long gone from Dodge County. I was hoping to spot Milton's name in a "survived by" list, but there was no mention of him. I did know that Milton was the son of William G. Humason, and fortunately William G. showed up in his father's obituary, allowing me to make some sense out of a confusing batch of names and dates.

Milton's grandfather, Lewis A. Humason, came to Rochester, Minnesota, from Ohio in 1860. Lewis may have been encouraged by another Humason family that had moved to the Rochester area a few years previously. If these two Humason families weren't related they soon would be: Lewis married Ellen Humason of the other Humason family. Ellen's brother Charles moved to Dodge Center in 1873 and opened a restaurant, perhaps inspired by the nearby Hubbell House. Lewis served in the Civil War and then returned to Minnesota and started a family, including William G., Milton's father. In 1878 Lewis moved his family to Visalia, California, where he spent the next decade in the milling business. In 1887 he moved back to Minnesota, to Dodge Center where Ellen's brother Charles was now a respected civic leader. Lewis built a grain mill there. We can guess that William G., after growing up in sunny California, was not charmed by Minnesota winters. William G. married a woman from California, who gave birth to Milton in Dodge Center, and before long they moved to Los Angeles, where William G. became a clerk in a bank. Ellen was in Los Angeles in 1905 visiting her son and grandson when Lewis died in Minnesota, prompting, according to Lewis's obituary, "her hurried and sad journey homeward." Lewis had spent his last seven years living in Mantorville, the county seat, where he served as clerk of the district court and was in charge of rebuilding the courthouse after a wind storm tore off its roof. His obituary said: "There is no man who can say ought against the character of L. A. Humason. Kind hearted and true to every friend or acquaintance and a cheerful word for one and all, the deceased will be greatly missed in Dodge County."

The year Milton's grandfather died was also the year Milton discovered Mt. Wilson, according to Gale Christianson's biography of Edwin Hubble. In 1905 there was no observatory atop Mt. Wilson, but there was a kids summer camp, which the fourteen-year-old Milton attended. Mt. Wilson looms over the Los Angeles basin, offering coolness, pine forests, wildlife, hiking trails, and winter snow that contrast with the desert environment below. Milton became so enthralled with the mountain that he was determined to stay. He got a job at the new Mount Wilson Hotel, washing dishes, carrying luggage, and attending to the mules that brought supplies up the mountain. This also meant that Milton's education stopped at the eighth grade. A few years later Milton became a mule driver, and now the supplies he was bringing up the mountain included construction supplies for the new observatory. Milton got to know the astronomers, especially the chief engineer, whose daughter he married. The observatory janitor's job allowed Milton to live in a rent-free cottage atop the magic mountain. Soon the magic included astronomy, which so fascinated Milton that he was asking endless questions of the astronomers. By all accounts, Milton was as good-natured as his grandfather. Milton's enthusiasm won the friendship of astronomer Harlow Shapley, who didn't take educational credentials so seriously, and Shapley got Humason started doing astronomical work. It would take longer for Edwin Hubble to accept Humason.

I headed for Dodge Center.

I was hoping to see a road sign welcoming me to the birthplace of Milton Humason, but there wasn't one. If I'd known where the Humasons had lived I might have seen the house that bore the human receiver of a new universe. Perhaps I passed it unknowingly. I had to settle for getting a feel for the two-block downtown. Many of its buildings were here when Milton was born. Most buildings were brick, but some were limestone, probably limestone from the still-active quarries near Mantorville, which also produced the limestone of the Hubbell House. I took a close look at the rough, pock-marked limestone and saw that it was packed with fossils, including some large plant forms. These fossils had offered to tell the young Milton that he was part of a vast, ancient story.

The Humason obituaries mentioned that Milton's grandparents were buried in the Riverside Cemetery. I went looking for the only trace of the Humasons that I would find here. I found the cemetery and headed for the older section. Some of the headstones were made of limestone, of fossils that now bore human identities, fossils that proclaimed the briefness and anonymity of all life forms even as humans forced them to proclaim the importance and imperishable memory of human lives.

I found the gravestones of Milton's grandparents: Lewis A. Humason, March 14, 1836-January 2, 1905. And Ellen, who had kissed Milton goodbye in Los Angeles and sadly hurried home to Lewis's large funeral. Had the preacher spoken the words "ashes to ashes, dust to dust"? Milton would help turn them into star ashes and stardust, ashes and dust that roamed through stars and space for eons and found its way to a thickening planet where one day it would rise up into human shape and act "kind hearted and true" and finally glimpse in fossil light the long, secret adventures of star ashes and stardust.

Going back to the Hubbell House for dinner, I paused outside to look at limestone fossils that told the latest chapters of the same story Edwin Hubble had found on Milton's photographic plates.

Carved into stone above one doorway was "Hubbell's Hotel." This doorway was no longer used, and in between its glass door and inner door were hundreds of box elder bugs, crawling and swarming. A spider had cast her net here; her dinner would be Hubbell Bugs. Another doorway, now walled-up, bore the chiseled words "Bar Room." Two stained-glass windows said "Hubbell House" in nearly faded paint. The windows were Tiffany style, with a centerpiece portrait of a lady's face being nuzzled by a dove.

I ordered the Hubbell Chicken Kiev.

Recipe for Hubbell Chicken Kiev (etc.): Why did the gas clouds cross the road? To get to the other side, the side with more gravity. Slowly turn up gravity until gas clouds condense into discs. Center of a disc will turn into new star, outer disc into planets. Almost all planets will spoil, nothing but rock or ice or gas. Only precise ingredients and temperature will produce suitable planet. Turn up volcanoes to outgas atmosphere. Toss in a touch of comets. Allow water to accumulate into oceans. Add lots of lightning bolts. Stir oceans a long time. Hope for the best.

The waitress set down a little loaf of bread, still hot and steamy from the oven, where it had performed an imitation of the expanding universe.

On the edge of Mantorville was a sign with a smiling comet passing Earth. It said, "Home of the Komets." The school mascot, with a 'K.'

The walls of the Hubbell House, with their displays of Gibson girl photos and flowery china, encouraged an image of elegance and order, but the people of Mantorville knew that they were not so far away from the primordial forces of nature, forces that were often chaotic. Farmers were reminded of this every year as they hoped for the right amounts of rain and sunlight; when farmers got their little loaf of Hubbell House bread they thought of this year's wheat harvest and whether they would keep the farm for another year. Back in 1858, only half a year after the Hubbell House opened with a grand Thanksgiving ball, a monster hail storm smashed 145 brand-new window panes, pouring hail onto floors and tables and ripping curtains. Hailstones were as large as twelve inches in circumference and weighed a pound. The storm lasted fifteen minutes and smashed windows and tore holes in roofs and destroyed crops for miles around. The hail killed chickens and dogs and bruised horses and cows, which ran hysterically. People ran into the Hubbell House and cowered in the roar of hail and thunder, in the eerie black-cloud gloom cut by lightning and by flashing visions of hail flying through windows.

For tonight, at least, nature was calm at the Hubbell House. But the humans were pretty lively. It was Saturday night, and the place was hopping. At the table next to me an elderly lady, perhaps in her 80s, was having her birthday party. It seemed that the Hubbell House gave out free cake and ice cream for birthdays. The lady's son handed her a present wrapped in a Harley-Davidson bag, but I couldn't see what it was.

For dessert I had the Hubbell Cheesecake, which had chocolate running off it like tiger stripes. Tiger, Tiger, tasting nice, on the forks of the night...in what distant deeps or skies burnt the fire of thine eyes? Speaking of philosophical poets, I wondered if Milton Humason had been named for John Milton. Perhaps to pass the long, cold, dark, lonely Minnesota winters Lewis Humason had read aloud from *Paradise Lost*, read of the creation and meaning of the world, read of transcendent mountaintops and punitive hailstorms, read:
Sing, Heavenly Muse, that on the secret top
Of Oreb or of Sinai didst inspire

That shepherd who first taught the chosen seed
In the beginning how the heavens and earth
Rose out of Chaos...

I returned to Mantorville the next morning. I would have eaten Hubbell Pancakes, but the Hubbell House wasn't open for breakfast. I stayed overnight in Rochester, in a wilted, century-old hotel right across the street from the main entrance of the Mayo Clinic. Some hotel guests stayed there waiting for the heart surgery of husbands who had eaten too many years of donuts. For breakfast the hotel offered nothing but donuts. I picked up a donut warily: it reminded me of the accretion disc of a black hole. Against my will, I was sucked in. At least the vast, many-block complex of the Mayo Clinic suggested that humans valued the lives we have been given.

The waitress who seated me for lunch explained that this was the chair where Roy Rogers used to sit. She pointed to the signature, "Roy Rogers and Trigger," among the famous guest signatures on the paper placemat. Roy used to come here for a nearby shooting match. I asked if Trigger came inside to eat, maybe a huge salad. She said that Trigger had to stay outside. One time a local character made a dinner reservation for two and showed up with his donkey, insisting that they had accepted the donkey's name. The donkey too had to stay outside. I observed that this was a good chair for a cowboy, as it was the only chair wedged into a corner, allowing Roy to guard his back against bad guys in black hats.

Sitting in Roy Rogers's chair, of course I had to order the Hubbell Hamburger.

Recipe for Hubbell Hamburger (and Hubbell Cheesecake): You'll need a cow. Start with an ocean full of organic molecules and hope for the best. Stir ocean until organic molecules, amazingly, form a cell. Allow cells to proliferate and fill oceans. Cells may need two or three billion years to form multi-celled organisms. Be very patient. Multi-celled organisms will sprout leaves, stems, legs, and heads, and eventually spread onto land and diversify greatly. Watch out for dinosaurs: if they threaten to become permanent, throw in big asteroid. Stir to encourage formation of grasslands and ungulates. You'll need one ungulate with ample milk and meat, and another ungulate to herd first ungulate and ride into sunset to tune of "Happy Trails to You" or "William Tell Overture."

My first reaction to the Hubbell Hamburger was that it was nothing special. But then I was judging it by its $7 price tag. When I nudged my mind back into Edwin Hubble's kitchen I saw the Hubbell Hamburger in a much larger context, saw that it contained immensities of time and work and talent, all of which created and nourished my own life.

I spent the afternoon wandering around Mantorville. I went up to the courthouse, with the roof Lewis Humason had been in charge of rebuilding. Lewis was also in charge of landscaping the courthouse grounds. I could only guess if his contributions were still visible after a century; if his work had included planting trees to replace trees damaged by the windstorm that tore off the roof, there were trees here that could be his work. When these trees started their photosynthetic careers, humans supposed that sunlight was generated by the gravity-induced burning of molten rock, which might keep the sun going for only twenty million years; but now I saw leaves aglow with nuclear fusion that had gone on for billions of years.

I checked out the ruins of a brewery built in 1874, ancient on the American timescale. Tourists came to Mantorville because it was a well-preserved late-nineteenth-century town, offering a sense of history and roots. Humans like to know where they came from. I made the rounds of historic homes and antique shops and gift shops, which survived off the traffic to the Hubbell House.

Two blocks from the Hubbell House was the Zumbro River. A red covered bridge, with a sign "Walk your horses, $2 fine," led to Goat Island, where goats once were corralled by the river. The opposite bank held limestone cliffs topped by autumn-colored trees. Downstream the river dropped over a low-head dam, whose slope created a broad, tall, glassy wave.

I settled onto a park bench beside the dam. I started reading an amusing book about Mantorville, but the river and its wave kept pulling on my attention. The wave was beautiful in its clearness, its perfect geometrical form, its sparking light. It seemed a permanent structure, but it was a structure of ceaseless motion. The wave was as hypnotic as a campfire. It was a flame made of water. No, something deeper than water. It was as if I was looking beyond the varied manifestations of energy and seeing the pure face of energy itself, the primal energy that ran the universe.

Every twenty or thirty seconds a leaf reached the top of the dam and spilled over it, accelerating down the wave front, disappearing in the froth below. There was a long parade of leaves coming down the river. Red leaves, orange leaves, yellow leaves, brown leaves. I watched leaves dropping off the trees atop the limestone cliffs and onto the water and heading downstream. Having just seen the water as primal energy, it as easy for me to see the leaves as galaxies, redshifted galaxies, streaming ever outward from the Big Bang. For hours the galaxies flowed toward me. Inside the galaxies, the galaxies turned into red leaves and dropped onto rivers and flowed away.

I was going to finish my Hubbell House weekend with a bang by ordering the Hubbell Prime Rib, but I discovered that it was served only on Saturday nights. This was just as well: I still deserved to be punished for having a donut for breakfast. I ordered the Hubbell Chef Salad, and the Hubbell Cheesecake for dessert.

Recipe for Hubble Cheesecake: You'll need a limestone restaurant. Compress leaves and shells of sea life to form thick stone. You'll need cheese makers: bring many settlers with Swiss names to Mantorville area. Finally, you'll need to evolve out of the rushing energies of the universe a being who can not only make cheesecake but taste in tiger-tiger cheesecake the sweetness and richness and strangeness of the whole universe, of the time and work and talent that goes into making life. You'll need to learn to perceive the Hubbell House not as a house of 1854 but as a house of fourteen billion years. You'll need to learn how to live in the house that Edwin Hubble built.

And it was good.

Lost in the Milky Way

I didn't realize that I was looking at the map upside down.

I was in the right spot, but I was looking in the wrong direction, so I was seeing nothing, nothing but a farmer's field, not the house I was looking for. I was trying to find the house where astronomer Harlow Shapley was born and grew up.

Historians of astronomy have called Harlow Shapley the Copernicus of the Milky Way galaxy, the man who mapped our galaxy and did for it what Copernicus did for the solar system: Shapley removed humans from an assumed center and placed Earth in orbit a long way from the center. Shapley also revealed our galaxy to be much larger than we had imagined, larger in size and in numbers of stars, billions of stars amid which our sun was quite average. Shapley also devised the method that enabled Edwin Hubble to establish that our galaxy was but one galaxy in a universe of billions of galaxies—one average galaxy in no special place in the universe.

My map was given to me by an elderly county historian, Bob, who drew a little dot where the Shapley house was, or where he remembered it, or where it had been. Bob said he hadn't been out there for many years, so he couldn't tell me if the house was still standing. Over twenty years ago Bob had photographed the Shapley house for history's sake, and he showed me the photos. Even then the house was long abandoned and in terrible shape, beyond being a haunted house; it was a ruin, a mere shell of warped, grey, rotting boards, of sagging roof and porch, of skull windows staring through moss and vines and the trees that had reclaimed the yard.

My truck stirred from the gravel road a long white limestone fossil cloud, a Milky Way swirling and dissipating. I slowed and stopped at the dot on my map, but there was nothing there, no ruin, no pile of wood, no foundation, only seamless rows of crops. Someone had obliterated the birthplace of a modern Copernicus. Perhaps this was done by a farmer who needed more space, or by a parent worried about his kids playing in a ruin. Old wood made good firewood, but it was a lot of work to dismantle a house and haul it away, and here in southern Missouri, where the winters were mild, there was little incentive to labor hard for firewood. Most likely someone simply had torched the house. The Shapley house became a brief supernova that poured carbon into the sky and onto the ground, carbon that was now rising into the architecture of plants, being set down on tables and eaten, becoming humans who did not guess that they were Copernicus houses.

It was as if the Shapley house never had been there at all. Indeed, as I was coasting down the road and looking at the map to find the Shapley family cemetery, I realized that the house really hadn't been there at all: I'd been looking at the map wrong. I turned around and went back to the dot and looked at the other side of the road. The Shapley house wasn't there either, just more crops.

As I looked more closely at this map, which located all the cemeteries and past and present towns of Barton County, Missouri, I noticed that whoever made this map seemed pretty confused about the area where the Shapleys had lived. The map included about twenty towns that no longer existed, towns that probably had been only country crossroads villages. One village had been plotted in the wrong place, about two miles from where it was supposed to be, and someone had added an arrow to point to its real location. This village was Duval, the Shapley's village. The map included about forty cemeteries, and two of them had been misplotted. The Mt. Carmel Cemetery had been corrected with an arrow pointing to its proper spot a mile away. The Shapley Cemetery had an arrow two miles long. It wasn't as if the mapmaker had misplaced Duval and the Shapley Cemetery through the same mistake, moving both of them two miles in the same direction. He had placed Duval two miles to the northwest of its proper spot, and the Shapley Cemetery two miles east, leaving Duval and the Shapley Cemetery about five miles apart, when, in reality, they were a mile and a half apart.

Assuming, of course, that the Shapley Cemetery was now plotted in the right spot. I began to have my doubts as I drove along the road nearest to it and didn't see anything that looked like a cemetery. I turned around and looked again. Atop a gentle ridge was a clump of trees that would make a good setting for a cemetery, but I couldn't see any gravestones or fence, and besides, this spot was too far from the road to fit the map. The only area that fit the map and wasn't farmland was thick with woods. Had the Shapley Cemetery been abandoned and overgrown?

Historians of astronomy have misplotted Harlow Shapley's birthplace as Nashville, Missouri. Nashville may be today's closest surviving town to the Shapley farm, but in Harlow's day the closest village was Duval, which had a general store, blacksmith shop, school, post office, and Presbyterian church. Today all that's left of Duval is one old house with a yard full of antique tractors, store signs, and gas pumps. Nashville isn't doing well either: its downtown is long abandoned and falling apart. Perhaps history readers who were a bit vague about the difference between Missouri and Tennessee imagined that Harlow Shapley was born in the glamorous capital city of country music.

Harlow Shapley mapped the Milky Way galaxy by using the first reliable measuring gauge for cosmic distances. Even as telescopes became much more powerful in the nineteenth century and astronomers made extensive photographic surveys of the sky, astronomers were at a loss to get a firm grip on the distances to the stars, and thus on the structure of the universe. There was no way to tell if nebulae were only gas clouds mingled with nearby stars, or if they were entire galaxies far outside our own. Using the parallax and trigonometry method that the Greeks had used to estimate the distance to the moon, astronomers figured out the distances to a few dozen of the nearest stars, revealing a startling, light-years-large universe. Beyond this, astronomers were stuck.

In 1912, two years before Harlow Shapley started work at the new Mt. Wilson Observatory, Henrietta Leavitt of Harvard Observatory proposed that one type of star, the Cepheid variable, might serve as a cosmic distance scale. The Cepheids showed a reliable cycle of growing brighter and then dimmer. Different Cepheids cycled at different rates, and it turned out that the rate at which a Cepheid cycled corresponded to its inherent brightness. The brighter a star, the slower it cycled. If you observed two stars cycling at the same rate this should mean that they had the same brightness. If one of those stars appeared only half as bright as the other, you could tell not only that it was father away, but exactly how far away. The Cepheids might be mileage signs on the highways of the sky.

In spite of some big doubts and problems in the Cepheid idea, Shapley ran with it, and he embraced the even more adventuresome assumptions that he could extrapolate from Cepheids to globular star clusters too far away to reveal Cepheids, and that the distribution of these globular clusters would correspond to the shape of our galaxy. Shapley spent several years observing Cepheids, producing a map with a large concentration of Cepheids a long way from Earth. Shapley declared that this concentration was the center of our galaxy and that our solar system was located about two-thirds of the way toward the galactic edge.

Shapley defended his findings at a 1920 meeting of the National Academy of Sciences, an event that has become so legendary that histories of astronomy call it "the Great Debate," though Shapley thought this term overblown. Shapley had an impish sense of humor and, in his autobiography, he claimed that Albert Einstein was present at the Great Debate and that during the long, boring preliminaries to the debate Einstein had turned to a neighbor and whispered, "I have just got a new theory of eternity."[1] In truth, Einstein wasn't even in America at the time, but prominent historians have repeated Shapley's joke as real history for decades now.

Shapley was right about the basic structure of the Milky Way galaxy, but he had overestimated its size, turning it from a supposed 25,000 light-years across to 300,000 light-years across, a dimension that seemed to make it hard for Andromeda and other nebulae to be galaxies similar to ours. Shapley's opponent in the Great Debate, Heber Curtis of Lick Observatory, believed that the nebulae were indeed other galaxies, and he tried to whittle Shapley's Milky Way back to 30,000 light-years across. In reality our galaxy is about 100,000 light-years across, and there's plenty of space for other galaxies. By deciding that our galaxy was the whole universe, Shapley removed himself from making some of the greatest discoveries in astronomy: the multitude of galaxies, the expanding universe, the Big Bang.

Shapley's eviction of humans from the center of the galaxy was widely discussed not just by scientists but by intellectual and cultural leaders. It might not have the same impact as Copernicus removing Earth from the center of the solar system and sending it flying, but still it was taken as a further demotion of humans from cosmic importance. Humans seem inclined to equate location with importance: there's great social status in having a house on the highest hill or the grandest street; numerous tribes define themselves as living at the center of the world; and New Yorkers look down on Missourians. Harlow Shapley was saying that humans lived on the astronomical equivalent of an average country road in Barton County, Missouri. And Shapley thought this was very good news, for Shapley hated human vanity.

Shapley felt that the long-running farce of human vanity wasn't funny anymore; now it was leading the human race to self-destruction. He saw World War One as a pointless waste of millions of lives for the sake of national vanity. He saw Hitler and fascism as pathological expressions of human vanity, now given racial form. After Hiroshima, Shapley was appalled that both America and Russia felt such self-importance that they were ready to destroy not only human civilization but perhaps four billion years of biological evolution. Shapley became deeply involved in humanitarian and peace work. When Hitler began persecuting Jewish scientists, Shapley became a crucial link in the lifeline that brought hundreds to safety in America. Shapley became a leading advocate for the United Nations and helped to found UNESCO. As Shapley became an increasingly visible spokesman for world peace, international cooperation, and American civil rights, the FBI tapped his phone, monitored his mail, and shadowed him. Joseph McCarthy accused Shapley of being a communist, and the House Un-American Activities

Committee hauled Shapley in and grilled him, but Shapley grilled them right back and became a hero for it. What Shapley's opponents didn't grasp was that in his political work he was serving the same vision he had served in his scientific work, a vision of human humility.

In the 1950s and 1960s Harlow Shapley was the most visible speaker and writer on things astronomical, and he frequently emphasized the humble place of humans in the universe and ridiculed human pomposity. He invoked his discovery of our place in the galaxy as giving him a special authority to speak about the philosophical place of humans. One of his books, whose very title promised a reverse-binocular view, bore the back-cover blurb: "*The View From a Distant Star* is a cosmic bird's-eye view of our insignificant planet in a solar system on the fringe of the galaxy. From this perspective man is a good deal less important than he thinks he is."[2]

Shapley scolded "the presumptuous primate" for his "egocentric vanities,"[3] "wrapped up in his self-esteem."[4] "They are crazy, I suppose, or at least they appear absurd, with their illusions and delusions. They have a game called 'for us the universe was made.' It is a song and dance with many variations. It is past belief how stuffy they are."[5] "The human race is not doing so well...It is beset with madness, with absurdities."[6] "He goes right on fighting thoughtless, futile wars, goes on behaving in a fashion more beastly than angelic."[7]

Fortunately, "correctives for our vanities are provided by modern science."[8]

"The displacement of the sun and earth from positional importance, the sudden relegating of man to the edge of one ordinary galaxy in an explorable universe of billions of galaxies—that humiliating (or inspiring) development is or should be the death knell of anthropocentrism. It should incite orienting thoughts by modern philosophers and theologians."[9]

"We have been deposed by the scientists from physical importance in the universe, and made ephemeral and peripheral. No longer can there be much esteem on the cosmic scale for a vain and strutting man-animal."[10]

"Short-term enterprises like our western civilization, or even the whole history of mankind, can be neglected as too momentary, too fleeting, for a clear recording in the cosmic panorama."[11]

"In view of the width of the cosmos and the slim hold on existence that Nature has provided for *Homo sapiens*, it would seem properly modest if we talked less about man being superior, less about him being the anointed of the gods."[12]

"Once we are free from the man-center illusion, our minds can roam over a universe that in size and power puts our inherited vanities to shame."[13]

Harlow Shapley's loathing of human vanity also meant that he loathed Edwin Hubble.

Edwin Hubble was also a rural Missouri boy, born eighty miles from Shapley, on the exact same latitude, and only four years later. Yet their personalities grew in very different directions.

Shapley maintained a down-home style and sense of humor, even when he became director of Harvard Observatory and was looked down on by the Boston blue bloods. In personality Shapley was similar to Harry Truman, who was also born in Barton County, only one year before Shapley. Missouri was first settled in the Jacksonian era and retained a populist ethos in which social pretensions actually counted against you, and thus Missouri has produced some of America's definitive personifications of populist spirit: Harry Truman in politics; Mark Twain in literature; Thomas Hart Benton in art. In his political views Shapley was a thorough populist, and perhaps it was no coincidence that his political activism thrived during the Truman presidency. In scolding humans for their cosmic pretensions, Shapley was pure Mark Twain.

By contrast, Edwin Hubble was similar to T. S. Eliot, who was born in Missouri within one year of Hubble. Eliot came from the St. Louis branch of a Boston blue-blood family, and he seemed to miss the old pretensions. T. S. Eliot went off to England and adopted a phony English accent and over-acted the role of an English gentleman. Edwin Hubble went off to Oxford to study law and never recovered. He came home with "a thick Oxford accent," wrote Shapley in his autobiography. "He was born in Missouri not far from where I was born and probably knew the Missouri tongue. But he spoke 'Oxford'...'Bah Jove' he would say, and other such expressions. He was quite picturesque."[14] Shapley mused that if you woke up Hubble in the middle of the night you would catch him speaking Missourian. Shapley and Hubble worked alongside one another at Mt. Wilson Observatory for several years, but "Hubble and I did not visit very much."[15] For someone who regarded World War One as proof of human insanity, it was galling that Hubble bragged about his heroic war career, expected to be called "the Major," and wore his military trench coat and boots to the telescope.

But what really galled Shapley was that he had held in his hands the key to the universe and he had dropped it and walked away to Harvard, leaving Edwin Hubble to pick up the key and become a greater astronomer than Shapley. The key was the Cepheid variable method

Shapley had used to map the Milky Way. Using Shapley's method and Shapley's telescope and Shapley's photographic plates for comparison, Hubble found Cepheids in the Andromeda nebula, thus proving that Andromeda really was a galaxy far outside our Milky Way. From there Hubble mapped out the galaxies, whose motions revealed an expanding universe.

Because Shapley couldn't see beyond the Milky Way, there wouldn't be a world-famous Shapley Space Telescope filling the world's newspapers with amazing photos, or a replica of the Shapley Space Telescope on the Barton County courthouse lawn. Historians wouldn't even think of preserving Shapley's childhood house. In the exhibits in the Barton County history museum in Lamar there's no trace of Shapley. You can find exhibits about Wyatt Earp, who started his lawman career in Lamar, and pretty shabbily too, fleeing town when the county charged him with fraud and embezzlement. Wyatt Earp fans come to Lamar and then continue their pilgrimage to Dodge City and Tombstone, where they get a big bang out of the OK Corral. Wyatt Earp pilgrims go visit Harry Truman's birthplace in Lamar partly because the Earps who remained in town bought and lived in the Truman house after the Trumans moved away. Bob the county historian told me that I was the only Harlow Shapley pilgrim he'd met. The Truman house is a Missouri state historic site, and within thirty miles of Shapley's birthplace is the birthplace of George Washington Carver, which is preserved as a national monument, and the birthplace of Mickey Mantle, who at least got the local baseball stadium named for him. It seems that discovering peanut butter and hitting home runs count for more than discovering the Milky Way galaxy. A third of a century after Shapley's death, no one has written his biography, and all his books are long out of print. Microsoft's spell checker happily recognizes the name "Hubble," but it thinks that the name "Shapley" must be a mistake.

Though Shapley scolded humans for their vanity, this didn't mean that Shapley himself wasn't vain. By all accounts he was a proud man. In its secret file on Shapley the FBI characterized him as "one of the supreme egotists of our time." Of course the FBI's definition of egoist was "he has an inherent dislike for authority."[16] But Shapley's pride was crushed by watching Edwin Hubble make the discoveries Shapley should have made. When Shapley wrote about how the multi-galaxy, Big Bang cosmos had deflated human pride, he was speaking from personal experience.

Shapley's insistence that humans are "ephemeral and peripheral" does seem to lend some poetic justice to Shapley being forgotten.

Shapley's philosophy of human insignificance was not entirely a product of his astronomical discoveries. Shapley came from a family of freethinkers. An 1889 history of Barton County, published when Harlow was only four years old, captured the Shapley family in a one-page sketch, mainly about Harlow's grandfather Calvin. Calvin Shapley was an anomaly in Barton County, an abolitionist and Union officer and Lincoln Republican in a county of Dixie-style Democrats. The county history book mentioned that Calvin kept trying to run for the state senate even though he never had a chance. The county history book usually listed people's religious affiliations, but in Calvin's case it said only: "He is a member of the Masonic fraternity, and his wife is a member of the Baptist Church."[17] Harlow Shapley wrote: "Our father also was without religion; the whole family has been without formal religion and has got along pretty well that way. I think that from the very first we were skeptical about the claims of religion and what it would do for you personally."[18] A century later a new county history seemed eager to present Harlow Shapley as a solid citizen, noting that he had attended the Duval Presbyterian Sunday school, making no mention of his atheism. Shapley said that he attended Sunday school simply as a social propriety.

There's no record of what the Christians of Barton County thought in 1913 when Harlow's father Willis was walking to his hay barn and was struck by lightning and killed, leaving him, in the words of the local newspaper, "burned to a crisp."[19] No doubt some Christians whispered that this was a predictable Old Testament punishment for atheists. Perhaps Harlow took it as a further proof of human insignificance. Harlow was in Paris on an astronomy tour when he got the news about his father, and "I wandered the Parisian streets for a long while in shock and blind mourning."[20] The city of lights; the spiraling boulevards; the Milky Way city. He could never bear to go back to Paris again. This was only one year before Harlow went to Mt. Wilson to map out the meaning of the city of lights in the sky.

If I had been able to find the Shapley family cemetery I would have found the grave of Harlow's father, the grave of a big bang that did not honor human life. But Willis Shapley too was lost.

The Shapley family's freethinking ways must have been reinforced when Barton County became the site of America's most famous experiment in atheism. From the pilgrims onward the settling of America has been rich with utopian communities. Utopias were usually founded with a religious mission, or sometimes from a philosophy of social and economic equality. One town was founded to be an atheist utopia. The town of Liberal, Missouri, was only a dozen miles from the Shapley homestead, and founded five years before Harlow Shapley was born.

The founder of Liberal was George H. Walser, a follower of Robert Ingersoll, "the Great Agnostic," who in the 1880s was at the peak of his influence. Large audiences packed Ingersoll's national lecture tours to hear him deny Christianity and proclaim the gospel of human reason and progress. To this day, streets in Liberal, Missouri, are named for Ingersoll and Darwin. George Walser was inspired by the idea that instead of being isolated and scorned as "the village atheist," atheists could have a place of their own, where freethinking would be honored.

Walser arrived in Lamar from Illinois at the end of the Civil War and became a successful lawyer and real estate dealer and got elected to the Missouri legislature. But the good citizens of Barton County didn't approve of Walser's complaints against Christianity. Walser bought some unclaimed land and took out nationwide advertising inviting atheists to move to Liberal: "the only town of its size in the United States without a priest, preacher, church, saloon, God, Jesus, hell, or devil."[21] Walser made churches illegal in Liberal, and required new arrivals to pledge never to hold religious services in town. Saloons too were illegal. "With one foot upon the neck of priestcraft," declared Walser, "and the other upon the rock of truth, we have thrown our banner to the breeze and challenge the world to produce a better cause for the devotion of man than that of a grand, noble and perfect humanity."[22] Hundreds of atheists answered the call. Walser founded Free Thought University, with seven teachers. He built Universal Mental Liberty Hall, which offered a Sunday school in which kids read from science books and performed chemistry experiments. Walser founded a 'Brotherhood' to carry out humanitarian work: "It presents opportunities for doing good which must engage the noblest impulses of the human breast...it leaves the supernatural to the speculation of those who find solace in pondering upon the unknowable and directs the mind of man to the wants of man."[23] And women too: a Ladies Progressive Lyceum was devoted to female equality. Liberal attracted nationwide attention.

Christian missionaries couldn't ignore Liberal and began moving into town and holding services. When Walser invoked the town's laws against churches, the Christians bought land on the edge of Liberal and started a rival town. Walser built a tall, barbed-wire fence across the border to keep the Christians out of Liberal. He posted guards at the Liberal railroad depot to tell passengers that if they were Christians they weren't welcome in Liberal. The railroad company took a dim view of Walser discouraging business and told him that either the guards and fence went, or they would close the depot. The depot stayed.

The downfall of Liberal came from Walser himself. Walser had been raised as a fierce Calvinist, and his rebellion against Calvinism fueled his atheism, but it seems that Walser retained a fascination with "the unknowable." The 1880s saw swelling interest in spiritualism, in ghosts, séances, mediums, and table rapping, and Walser got caught up in the enthusiasm and took many of his followers with him. Committed spiritualists began moving to Liberal. Soon Liberal had a Spiritualist Hall, which hosted some of the famous mediums of the time. In the final séance in Spiritualist Hall people wrote questions on a chalk board and hoisted it—by rope—to "the spirit world," from which it was lowered with true and brilliant answers. Then a fire broke out and two local men fled from a secret door in the ceiling, chalk still in hand. Walser built an elaborate park for national camp meetings of spiritualists. Walser designed the Liberal cemetery as a giant circle where everyone would be buried with their feet pointing at the center, the center where Walser would be buried—thus at the great resurrection of spirits all his followers would be facing him.

To unrepentant atheists the spiritualist goings-on in Liberal were outrageous. With the atheist community divided and not too impressed by "grand, noble and perfect humanity," the Christians continued making progress in Liberal. The Universal Mental Liberty Hall was converted into a Methodist Church, which soon afterward was struck by lightning and seriously damaged. At long last, the Christians legalized saloons.

In his autobiography Harlow Shapley makes no mention of Walser or Liberal, but the Shapleys must have taken an interest in the atheist utopia/circus only a dozen miles away. It's especially intriguing whether Harlow was influenced by O. E. Harmon, who wrote the history of Liberal. Harmon was deeply interested in astronomy and had made eclipse calculations that impressed professional astronomers. Harmon had wanted to become an astronomer, but health problems discouraged this ambition and prompted him to settle in Liberal in 1897, when Harlow was twelve. Harmon wrote for *Popular Astronomy*, the leading astronomy magazine of the time, including an article on "The Astronomy in Shakespeare." Harlow Shapley's first published article was in *Popular Astronomy* in 1909: "Astronomy in Horace." In any case, it probably wasn't a coincidence that one of the most visible atheists of the twentieth century grew up in the midst of the most famous atheist experiment of the nineteenth century. The failure of Liberal may have contributed to Shapley's disdain for everything spiritualistic and psychic.

In his popular writings Shapley was circumspect about his atheism and would confine himself to general ridicule of human vanity. But sometimes, especially when provoked by the prospect of nuclear annihilation, Shapley became more explicit about religion, especially as an ethical system: "We need an ethical system," he wrote in his introduction to *Science Ponders Religion*, a book on the inadequacies of religion that he edited in 1960, "suitable for now—for this atomic age— rather than for the human society of two thousand years ago...If atomic war tools are available to angry and vain and stupid men, and are used— then a grim final curtain will close the human play on this planet. It will be truly a judgment day—a day of our own *bad* judgment."[24] In such a mood Shapley would openly call for the abandonment of too-small, too-righteous gods. It was Shapley's atheism that made Joseph McCarthy positive that Shapley must be a communist.

I cruised past the Shapley homestead again, hoping that from a different angle some structure would pop out. I was also looking for a sign of the Shapley School, a one-room school on land donated by Harlow's grandfather Calvin, which Harlow attended for awhile in a class of three taught by his sister. Bob the county historian told me that the school was across the road from the Shapley house. In his autobiography Shapley left the impression that it was farther off. I couldn't find it anywhere, only a trace of a driveway that ended at a lone, tall tree. I was especially disappointed to find no trace of a small, stone house, of which Bob had shown me photos. He thought that this was the first house Calvin Shapley had built, near the later, two-story wooden house. No one could have casually burned down the stone house; someone must have disassembled it. The stones must be around here somewhere, in a farmer's fence or ditch. By now the Shapley past of these stones was forgotten.

The ants, no doubt, were still there. Harlow Shapley became fascinated by ants at an early age, and they helped interest him in the natural world: "There was no instruction in natural history in the country schools I attended. I turned over rocks and saw some little crawling things called ants—I used to look to see if they had guests with them, such as beetles."[25] The rocks Shapley turned over and the descendants of the ants he studied were probably still there.

Shapley even published scientific papers on ants. He studied ants atop Mt. Wilson by day as he studied the Milky Way by night. In looking at a line of ants moving along a concrete wall he noticed that they moved faster in sunlight, slower in shadows. He set up "speed traps" and used "a thermometer and a barometer and a hydrometer and all those 'ometers,' and a stop watch" to prove that ant speed was a function of air temperature and no other factor. "By observing half a dozen ants through my speed trap, I could tell the temperature to within one degree."[26] The ants were just like Cepheid variables, their cycle rates a function of heat. When Shapley's train broke down on his way to the Great Debate, Shapley wandered the tracks looking at ants.

Ants showed up quite a bit in Shapley's popular writings, especially when he wanted to scold humans for their vanity and make them feel smaller. You can tell that Shapely was tempted to say that humans are mere ants swarming obliviously upon the earth, but he never does, and in fact he says he would consider this to be either an insult to ants or a compliment to humans. Ants, after all, have survived for many millions of years, and their societies are logically organized and don't self-destruct from ego and foolishness. Shapley offered the success of the ants as another measure by which humans should feel humble: "The social Harvester ants provided, for me at least...a vantage point from which to view the current evolutionary troubles of our multi-nation society."[27]

Shapley also used extraterrestrials for the same purpose he used ants. Shapley was one of the first astronomers to argue on twentieth-century scientific grounds that the universe must be rich with planets and biological evolution and intelligent life. He invited us to measure ourselves by a cosmos with millions of intelligent species, who had survived for ages by being peaceful and wise.

Just when Shapley starts sounding like a misanthrope, he inverts his binoculars to an expansive view and explains that his real motive is to wake up humans to the wonder of the universe and our presence in it: "The universe...is much more glorious than the prophets of old reported, and we are actors in a greater show than the old billing led us to expect."[28]

I was born in the town where Shapley learned astronomy, and for many years I lived two blocks from the site of the observatory he used.

Shapley came to the University of Missouri in 1907, intending to study journalism. At age fifteen Shapley had taken a job as a reporter for a humble, small-town newspaper near home. He covered car crashes and petty crimes and egomaniacal politicians, and one time he proved how a mathematical horse at a circus was really a fraud, but it turned out that his editor only wanted to blackmail the circus into buying advertising and never printed Shapley's expose. At this time Joseph Pulitzer, who owned the *St. Louis Post-Dispatch*, was promoting the idea that journalism should be a respected profession with high standards and a credo of public service, and in response the University of Missouri established the world's first school of journalism. Pulitzer's idealism appealed to Shapley, so he headed for the university, only to find that the journalism school wouldn't open for another year. To fill the time Shapley took an astronomy course. He later claimed that he owed his astronomy career to astronomy being the first subject in the college catalog that he knew how to pronounce, after the puzzling 'archaeology'. This was another Shapley joke that some historians have repeated seriously. If Shapley had been serious, he would have gone into accounting or agriculture.

The university's seven-inch telescope had been acquired from a failing Kentucky college, and the rest of its equipment was modest at best. "We had a very bad photometer—I would swear at it, even now, if I knew how to swear at photometers."[29] But Shapley had a first-rate professor, Frederick Seares, who was studying variable stars and who set Shapley to work on them. Seares soon moved to Mt. Wilson Observatory and would open its doors to Shapely. Through such coincidences, hack reporters are turned into great astronomers, and universes are remade. Seares always thought of Shapley as a journalist and encouraged his popular science writing.

When Shapley's would-be school of journalism needed a site for its own building, the university disassembled the observatory and rebuilt it on the edge of campus. The journalism school would produce some prominent science journalists, who would even write about astronomy, but they probably never realized that they were studying on the very spot where Harlow Shapley had learned astronomy. There are various historical markers around campus, but there's no trace of Shapley. There's even a statue of Beetle Bailey, who was created there. The observatory was neglected for years until most of its instruments were broken and a leaking dome was rotting the wooden tube of the telescope Shapley had used. Fertilizer stored in the basement made the building stink. I still remember how decrepit the observatory looked in the early 1960s. Finally it was torn down. It had stood near a huge landfill that was creating a parking lot for a planned basketball arena, and I assume that the observatory was tossed straight into the landfill. The old telescope was saved, but it disappeared into a storage basement.

I made one more try at coasting down the road to try to spot the Shapley Cemetery. There was a car at a nearby farmhouse and I thought of stopping and asking directions and permission to traverse their field. But how would I explain why I wanted to see the family graves—or anything at all—of a man whom everyone else had forgotten? I finally decided that maybe what I was supposed to find was exactly what I had already found—nothing at all. Maybe it was only appropriate that the prophet of human insignificance should have disappeared without a trace. I drove off.

Still, I knew that for Harlow Shapley getting lost was merely the risk that humans had to take for the rewards of being part of a huge, ancient, magnificent cosmos. For Shapley this risk was well worth taking: "We are actors in a great cosmic play where the performers include the atoms, the galaxies."[30] "With our confreres on distant planets, with our fellow animals and plants of the Earth's land, sea and air, with the rocks and waters of all planetary crusts, with the photons and atoms that make up the stars—with all these we are associated in an existence and an evolution that inspire respect and deep reverence."[31]

What Shapley was afraid of was that humans would refuse this risk and hide from the universe in smug little shells that protected our vanity. Shapley knew better than anyone the hesitation of humans to embrace the universe, for he had grown up a dozen miles from a town dedicated to living in the natural world yet which had lost its vision and nerve and retreated into spiritualism, into a universe where humans would never be lost and forgotten since humans became spirits who could rap on tables and relay messages through mediums. Shapley feared that humans would rather see a chalk message from your father's ghost than see the chalk itself as a message from the fossil and geological past.

Yet if humans could be lured out of their vanity shells to live in the shells of Earth and galaxy and cosmic time, they would not only have a better chance of surviving as a species, they could rise to a great role as actors in the cosmic drama. They could honor eons of evolution not just by not destroying it, but by openly celebrating it. They could take a starring role in a galaxy whose stars could never so shine so lucidly as sentient life. They could even serve as mediums through which the dead could speak, the dead of all the biological eons, the dead now of limestone cliffs, even the dead who had never lived: the dead planets, the dead gas clouds, the dead stars, the dead Milky Way galaxy itself. Through its sentient life the long-lost Milky Way galaxy could now recognize itself and its journey and its beauty and proclaim itself to be a home at last.

You Can't Go Home Again

One night I tried to touch my past, but the past was not there.

For years I lived close to the playground where I had played as a kid, and occasionally I walked past it. One night, for reasons known only to my neurons, I paused and wandered out into it. I walked over the ground where once I had run. I stood where once I had stood with a baseball glove and expectantly awaited the next pitch. I thought of the hundreds of victories and defeats I had experienced here—but I found I could taste almost nothing of the pleasure or frustration they had contained. I thought of the friends who had played here with me. For a moment I saw us playing together, with no thought of tomorrow, and I heard our yells, but then I was alone in the silence and the night again, knowing that those friends were grown and scattered and would never again share a baseball game. I was alone, that is, except for two cars, for now the playground was a parking lot. I imagined that the owners of the cars were somewhere sweeping floors and emptying wastebaskets, erasing small pieces of other people's yesterdays.

After awhile I left the playground, left it to people who would see in it only a parking lot, and I headed home. But a couple of blocks from home I paused again, at another place where I could look into the past. This was not my past, but the past of Harlow Shapley. I stood next to the site of the observatory where Shapley had learned astronomy. I imagined young Shapley standing here and gazing through a telescope, not suspecting that the work he was beginning here would eventually lead him to make great discoveries about the universe. But then I was back in the present, and instead of an observatory I was seeing one of the buildings of the University of Missouri School of Journalism.

I started to leave, but then a thought struck me. I looked again at the building before me, and I realized that this was not really the site of Harlow Shapley's observatory. Nor was the parking lot in which I had been standing really the site of my childhood playground. The work of Harlow Shapley, which had led to the discovery of the expanding universe, now led me to realize that the place where Shapley had learned astronomy and the place where I had played as a kid were really countless billions of miles away, somewhere far in space and growing farther every second.

I gazed up into the night, into the stars we sometimes mistake to be changeless, and I saw them rushing through space like grains of sand riding the winds of a hurricane. I saw myself riding on a planet that was flying around the center of a galaxy, and riding in a galaxy that was flying away from the Big Bang. I felt my heart beating, and I realized that each heartbeat was occurring many thousands of miles away from the last one. My heart beat, and I watched the point where it occurred streaking away into space as the planet left it behind. My heart beat again, and in only moments the point where it occurred was farther away than the moon. Gazing far beyond the moon, I saw the long trail of heartbeats I had laid through space, a long trail that was my life. I saw all the places and all the actions of my life stretched out one behind the other. I saw the spot where I awoke this morning, a spot that now was only empty space. I saw the spot where last week I sat and felt upon my face the warmth of the sun, a spot that now was far behind the sun and utterly cold. I saw meals that had stretched across millions of miles, and books that had kept me company for many billions of miles. Somewhere far, far back I saw myself standing in a playground and throwing a baseball a thousand miles. I saw myself gradually growing smaller until I was a baby and then was not there at all. Surrounding the line of my life were the lines of billions of other lives, all twisted together like the threads of a massive rope, a rope that Earth lays across the universe, the rope of human history.

I looked back along that rope and saw the places where other people had lived their lives. I saw where Edwin Hubble had sat at a desk and calculated how fast the universe was expanding. I saw where Albert Einstein had invented an escape-clause equation to avoid the absurd idea that the universe must be expanding. I saw where young Harlow Shapley had first looked through a telescope, not suspecting that he was flying around the galaxy and away from the Big Bang, but beginning the work that would reveal it. I saw where Thomas Edison had invented the light bulb, a place that now was very dark. I saw where Beethoven had conducted one of his symphonies, a place that now was very silent and still. I saw where Roman legions had fought fierce battles, but there were no monuments marking those battlefields. I saw where a Greek philosopher had wondered whether it just might be possible that Earth was not the fixed center of the universe. I saw the vast distance over which humans had built Stonehenge. I saw where humans had huddled inside caves and gazed out into the darkness. I saw where some vaguely human creatures had wandered over African savannas.

Then I looked beyond human history and saw the history of Earth stretched through space in a long helix. As Earth circles the center of the galaxy, the path Earth traces through space weaves back and forth, forming a helix. I looked along this helix and saw some tinier helixes gradually changing; I saw the life of Earth flowing from one shape into another. As I looked farther and farther back I saw the dinosaurs shrinking into smaller reptiles, which turned into fish and slithered back into the sea. I saw plants growing smaller and simpler and withdrawing from the land until it was empty. I saw fish growing smaller and simpler until they were no longer fish but tiny creatures riding helplessly in the waters. I saw these creatures losing more and more of their cells until they consisted of only a single cell, and I saw these cells growing cruder until they fell apart into molecules. I looked into the ages before life and saw where rain had pounded the land and turned into rivers and oceans, and where volcanoes had poured out lava and black clouds. I saw where meteors had bombarded the newborn planet, and then I saw the planet forming out of a dense, swarming cloud of gas and dust. Earth had traced a path stretching a third of the way back to the Big Bang. Before Earth was formed, I saw the threads of the future Earth weaving through several generations of stars. I saw our galaxy emerging out of formlessness. Finally at the start of it all I saw the Big Bang.

I stood there awhile and watched the whole history of the cosmos receding into the darkness. I watched all the places of my life racing into cosmic history, receding and receding and receding.

Then I turned and started walking, walking on pavement I sometimes mistook to be steady but now felt to be falling and falling; walking on a giant bullet that was flying through space, flying from a gun called the Big Bang. I turned and headed—no, not home, not to a cozy place I returned to again and again, not to a place where much of my past had been lived, not to a place where each day I added a little more life to all I had lived there before, for no person on Earth and no being in the universe has such a home. We are all homeless; we are all cosmic refugees forever wandering, lugging with us little scraps of our past. I headed not home, but merely to the rooms where I did most of my wandering.

The Music of the Spheres

"Like Henrietta Leavitt, Cannon was deaf...The following week she and a
friend invited Shapley over for dinner and they all went to the symphony."
—George Johnson, *Miss Leavitt's Stars*

Yes? And what kind of symphonies do deaf astronomers enjoy?

The Cannon in question was Annie Cannon. She and Henrietta
Leavitt worked together at Harvard Observatory early in the twentieth
century, under director Harlow Shapley. Even today, when women find
open doors to careers in astronomy, Cannon and Leavitt remain two of
the most famous women astronomers in history. Curiously, both were
deaf.

I like to imagine Henrietta Leavitt and Annie Cannon sitting
together in a concert hall, perhaps with Harlow Shapley sitting between
them, watching the performers moving their bows back and forth on the
strings, moving their fingers up and down on the horns. Barely moving
anything else. I've always wondered why concert audiences sit there for
two hours watching motions so undramatic. The guys who pound on
the drums and crash the cymbals can be pretty dramatic, but you usually
have to wait a long time for them to have their brief turn. If you look
around a concert hall you can usually see some audience members with
their eyes closed, and more people would like to close their eyes but
they are afraid of being mistaken for being asleep. Surely they have the
right idea. They are giving all their attention to their hearing. If there are
any blind people in the audience, they may be hearing the music better
than anyone. When humans lose one of their senses, other senses often
become more acute. This isn't just a matter of attention either, but of
physiology, of neural compensation. Deaf people sometimes develop a
keener sense of vision. Perhaps Henrietta Leavitt and Annie Cannon saw
the performers' hand motions more keenly than anyone else. Perhaps
they were fascinated by the same process they saw in a weaver's hands
playing a loom: the emergence of meaningful patterns out of numerous
little events. In the same way, cosmologies emerged out of thousands
of photographic light-notes. Perhaps they were living in a world where
sound just didn't matter very much. In space, no one can hear the
symphony.

It's hard to find much reliable information about the deafness of
Annie Cannon and Henrietta Leavitt, or much else about them. Both
have faded into biographical silence. When science journalist George
Johnson was researching the book quoted above, he gave up trying to

make it the first biography of Henrietta Leavitt, for he could find only scraps of information about her. He also found that much of what had been published about her was wrong. One book about deaf scientists suggested that Leavitt was deaf by the time she enrolled in Oberlin College, but Johnson found that Leavitt had enrolled in Oberlin's conservatory of music. Many sources repeated the statement that Leavitt was "extremely deaf" a few years later when she entered Radcliffe, but Johnson found a letter from over a decade later in which Leavitt expressed concern about her faltering hearing. Annie Cannon lost her hearing while in college, but even those who worked with her left different impressions about the extent of her deafness. Harlow Shapley said that Cannon's hearing was "pretty much lost," while Cecilia Payne-Gaposchkin, who would benefit from the careers of Cannon and Leavitt by becoming the first female chair of Harvard's astronomy department, reported that Cannon wore a hearing aid, which she would remove when she wanted to be undistracted. This leaves open the possibility that with her hearing aid Annie Cannon could have had some sort of musical experience. If the gentle string music was lost on her, perhaps music with loud horns, drums, and organ could have gotten through. *Also Sprach Zarathustra* would have been perfect. Did Annie close her eyes to concentrate and see the blackness and imagine this to be the blackness of space and the music of space?

Annie Cannon and Henrietta Leavitt took part in the most ambitious sky survey yet attempted, a census seeking to catalog every observable star according to location, brightness, and color. Cannon and Leavitt and a staff of other women, called 'computers,' studied tens of thousands of photographic plates and measured every dot of light. They were paid as if this was clerical work, but it gave them the chance to be the first to notice important patterns. Cannon noticed that stars belonged to distinct types and devised a classification system still used today, a system that pointed to the stellar lifecycles astronomers would later discover. Leavitt was studying Cepheid variable stars, which vary in brightness, when she noticed that the brighter a star was, the longer its cycle of variation lasted. Leavitt published a paper about this period/brightness law, and Harlow Shapley, then at Mt. Wilson Observatory, realized that this could be the key to mapping the universe. Astronomers had never had any reliable gauge for determining cosmic distances. You could never tell how far away a star was because its brightness could be because of distance, or because stars varied greatly in intrinsic brightness, or because interstellar gas and dust made stars appear dimmer. But with Cepheids if two stars with identical periods were different in apparent brightness, it was definitely due to distance, and you could calculate exactly how much distance. Shapley wrote to Harvard Observatory's director William Pickering: "Her discovery of the relation of period to brightness is destined to be one of the most significant results of stellar astronomy, I believe."[32] Shapley used Cepheid variables to map our galaxy. Edwin Hubble used Cepheids to prove that there were galaxies outside our own, and to start to prove that we were part of an expanding universe, expanding from a Big Bang. Shapley's work with Cepheids won him the directorship of Harvard Observatory, where he became Henrietta Leavitt's boss, at least for the final year of her life. As Leavitt lay dying of cancer, Shapley visited her and expressed his appreciation for her work.

On his first day at Harvard Observatory Shapley started appreciating Annie Cannon's "phenomenal memory." He told her that he wanted to see a photographic plate with a particular star on it, and she knew the five-digit plate catalog number from memory, out of many thousands of plates. Cecilia Payne-Gaposchkin decided that Cannon possessed an uncanny visual sense that served not only as a photographic memory but as an ability to see subtle details where others saw almost nothing. In looking at Cannon's photographic plates Payne-Gaposchkin felt:

> It seemed impossible that anyone could see enough in those tiny smears to classify the spectra...In the last years of Miss Cannon's life, Henry Norris Russell used to say 'Someone ought to find out from Miss Cannon exactly how she classifies each spectral type.' I argued with him that she would not be able to tell them because *she did not know.* She was like a person with a phenomenal memory for faces. She had amazing visual recall, but it was not based on reasoning. She did not think about the spectra as she classified them—she simply recognized them.[33]

Cannon's visual powers were probably due to her deafness. Shapley recognized another way in which Cannon's deafness had encouraged her astronomical talents: "Miss Cannon had the disadvantage—or advantage—of having had some sort of infection while she was in college, and as a result her hearing was pretty much lost. That handicap took her out of the social life and put her into science. In some ways I feel rather grateful to that particular bug."[34]

I don't want to leave Shapley sounding insensitive about Cannon's deafness, so it should be noted that quite a few prominent deaf people have indeed considered their deafness to convey advantages. Thomas Edison almost celebrated his deafness for enhancing his powers of concentration. Konstantin Tsiolkovsky, the Russian pioneer of rocketry, found social life so awkward that "this caused me to withdraw deep within myself." "My deafness made me ridiculous. It estranged me from others and compelled me, out of boredom, to read, concentrate, and daydream."[35] Jean-Jacque Rosseau's partial deafness encouraged him to shun society and embrace nature, an orientation he codified into Romantic philosophy. Gilbert White, in *The Natural History of Selborne*, bemoaned a deafness that prevented him from hearing the birds singing, but it also probably encouraged the immersion in nature that made White an inspiration for generations of naturalists. We don't know how Cannon and Leavitt felt about their deafness, but they certainly didn't miss the birds singing in space.

It was an odd coincidence that two deaf women not only worked at the same observatory at the same time but made important discoveries there. Leavitt didn't owe her job to Cannon's precedence or mentoring. Leavitt started her association with Harvard Observatory as a volunteer at age twenty-five, before she lost her hearing. In an even odder coincidence, the English astronomer who had first discovered Cepheid variables in 1784 was also deaf.

John Goodricke was born deaf but was enrolled in one of the first institutions devoted to educating the deaf, Braidwood Academy. One day when Goodricke was nine years old, Samuel Johnson, who was partially deaf, visited Braidwood and was amazed to see deaf children doing challenging math. Goodricke developed acute visual sensitivity. As a teenager Goodricke met astronomer Edward Pigott, who was interested in variable stars. Pigott put Goodricke to work observing the variable star Algol, and Goodricke conceived the idea that Algol varied because it was actually a system of two stars, with a dimmer star regularly eclipsing a brighter star. This concept explained a large proportion of variable stars, but not all of them. Other astronomers had never even noticed that the star Delta Cephei was variable, but Goodricke not only noticed this, he found that Delta Cephei varied in a pattern that couldn't be explained by the concept of eclipsing binary stars. Delta Cephei had to vary because some internal process actually made its brightness fluctuate. Goodricke found more stars that fit this pattern and called them Cepheid variables. Goodricke's discoveries won him membership in the Royal Society at age twenty-one, but only two weeks later Goodricke died of pneumonia.

Adding still more to the connection between deaf astronomers and variable stars was Robert Grant Aitken, who spent forty years at Lick Observatory in California. Aitken was once injured by a car on a Berkeley street because he couldn't hear its approach or horn. At the same time that Henrietta Leavitt was working on Cepheid variables, Aitken was preparing his 1918 book *Binary Stars* based on his study of

3,000 binary stars. Aitken, Leavitt, and Cannon each got moon craters named in their honor. Moon craters like giant ears listening to the silence of space.

The many connections between deaf astronomers and variable stars might leave you thinking that there must be many deaf astronomers, but in fact this essentially completes the list of prominent ones. We are invited to indulge in the notion that variable stars must have some special attraction for deaf astronomers. Variable stars are all about rhythm. Rhythm is the backbone of music. Variable stars are playing a silent music, a music of long vibrating strings of light, of gravity horns, a music that contains a far deeper structure and harmony than merely human music. In finding the relationship between Cepheid periods and luminosity, Henrietta Leavitt had found the same kind of law that inspired Pythagoras to proclaim that the cosmos was filled with the music of the spheres.

The Greeks became enthralled by the idea that behind the confusing phenomena of the world lay a hidden order, a natural order. Thales suggested that the whole cosmos was made of water, which took on various forms and actions. Other Greek philosophers tried to build the world out of four elements: earth, air, fire, and water. Music gave Pythagoras a revelation that the deepest order of the cosmos was mathematical. Legend says that Pythagoras was walking past a blacksmith's shop when he noticed the varied sounds of the hammers striking anvils. He went inside and studied the hammers and realized that their weights had an exact mathematical correlation with the musical intervals they produced. Pythagoras then went home and stretched out gut-strings and plucked them and heard them announcing mathematical ratios. If music was math and if the sky showed mathematical order, then the sky must be musical. For Pythagoras the sky consisted of giant crystal spheres one inside another, revolving spheres with holes that allowed you to see the cosmic fire beyond, see it as stars and planets. The stars and planets moved in reliable ratios that were just like musical intervals. This must mean that the crystal spheres were musical, giving out great tones, tones adding up to a vast harmony. The cosmos was like a giant lyre playing beautiful music.

Pythagoras's vision would cast a long spell. Plato turned the music of the spheres into a math-obsessed Demiurge who created a cosmos of math-filled order. Cicero turned the music of the spheres into popular literature in *Scipio's Dream*. Early Christians like Clement of Alexandria used Pythagorean imagery but replaced the Demiurge with Christ. Tycho Brahe incorporated Pythagorean ratios into the architecture of his mansion-observatory. Kepler became so obsessed with the idea that the solar system displays geometrical perfection that he ignored his own discovery that the planets move not in circles but in ellipses, and in his *The Harmony of the Universe* he indulged in a Pythagorean quest for the musical codes in the sky. Newton believed that he had merely rediscovered knowledge that Pythagoras had kept obscured within his mystery cult. Mozart turned *Scipio's Dream* into *The Magic Flute*, which sings of "the unwonted harmony of the moving spheres you hear."

Could it be that if Annie Cannon and Henrietta Leavitt and Harlow Shapley went to a musical performance together, they went to *The Magic Flute*? Then they would have heard Scipio asking why, if the heavens create such magnificent music, humans are deaf to it. The goddess Costanza replies, echoing Aristotle's answer:

> It too far exceeds the perception of your senses.
> The eye that turns upon the sun
> cannot see the sun at which it gazes,
> dazzled by that same
> excess of splendor.

> He who lives on the banks
> of the tumbling waters of the Nile
> does not notice the noise
> of the raging torrent.

Compared with not being able to hear the music of the spheres, not being able to hear a symphony or opera could have been too trivial for Annie and Henrietta to worry about. Or do these words imply that someone with especially keen senses might have a chance of hearing the music of the spheres? Annie and Henrietta certainly found subtle harmonies that would have impressed Pythagoras. If humans require a noise much louder than the Nile, it was Henrietta Leavitt who first creaked open the lid of the Big Bang.

After Kepler the crystal spheres and their music faded out of the scientific cosmos, but Pythagoras's vision of the mathematical order of the cosmos became one of the foundations of science. The universe was built of sphere within sphere of extraordinary order, of atom within cell within body within ecosystem within planet within solar system within galaxy. In every generation astronomers have experienced Pythagorean awe at seeing the secrets of the universe unfold in their equations. Albert Einstein, who preferred playing Bach and Mozart on his violin because of the mathematical order in their music, said that mathematical elegance was one of the first things he sought in a scientific theory, one of the surest signs that a theory must be true. Once, the night sky was built out of stories and poetry, but now astronomy textbooks were full of equations, which were poetry of a different kind, a deeper kind in which "cosmos" rhymes with "life," in which human life is articulated by a vast sequence of rhymes.

Annie Cannon and Henrietta Leavitt must have felt some Pythagorean awe that a few numbers could make sense out of their years of searching the stars, and perhaps this makes it a bit less likely that if they went to the concert hall with Harlow Shapley they went to see Gustav Holst's *The Planets*. Holst's work was based on a different vision of the cosmos, an astrological cosmos in which planets were gods whose personalities and eccentric whims ruled human life. Even Shapley would have been deaf to such music.

It would have been more appropriate for them to go hear music by William Herschel. William Herschel spent most of the 1760s composing music for orchestra and solo instruments, especially the organ. Herschel was the organist at a chapel in the resort town of Bath, England. His father had been a musician in a German military band, but William didn't care for military life or, it seems, military music, and he moved to England to pursue a career in loftier music. In addition to playing the organ he gave music lessons and conducted choirs. Herschel probably heard the mathematical order in music, for he enjoyed studying mathematics. Herschel was thirty-five years old when he became captivated by astronomy, and it wasn't the sky that captivated him but a book, which suggests that what captivated him was ideas about cosmic order. Herschel began building ambitious telescopes and observing the sky. Sometimes he went straight from performing music to a night of observing, or straight from the telescope to the Sunday morning organ. Perhaps the music and the stars blended thoroughly in his mind. Certainly he heard the music of the spheres in 1781 when he spotted a sphere changing position from night to night. He was the first astronomer to discover a new planet, Uranus. By the way, William Herschel's sister Caroline, who gave up a singing career for astronomy, became one of the few women astronomers in the same league as Annie Cannon and Henrietta Leavitt. Yet in the end it's very unlikely that Cannon, Leavitt, and Shapley would have had the chance to hear Herschel's music, which history has allowed to go silent.

The years that Cannon and Leavitt worked together were the heyday of the silent film. Silent films placed deaf people and hearing people on equal terms. A bell ringing on a movie screen was equally silent for everyone. Even the greatest of bells, the bells of Notre Dame Cathedral, were silent in *The Hunchback of Norte Dame*. The deaf bell ringer,

Quasimodo, was played by Lon Chaney, whose real-life parents were both deaf. The bells of Notre Dame were supposed to be the voice of God summoning humans to worship him, but their loudness had rendered Quasimodo deaf, and no one in the movie audience heard God's voice either. We can imagine Cannon and Leavitt and Shapley going to silent movies together and watching church bells proclaiming nothing, watching human mouths moving without meaning anything; yet when the camera glimpsed the night sky, the sky was just as right as ever: the stars were tiny Pythagorean bells.

It was to summon the deaf to church that a Spanish monk, Pedro Ponce de Leon, became around 1550 the first person to teach language to the deaf. Ponce de Leon worried that if the deaf couldn't understand the word of God or make confession, their souls would be lost. Christianity placed unprecedented importance upon The Word. When the first Greek Christians replaced the Demiurge with Christ, Christ inherited the Demiurge's Logos, the rational order at the heart of the cosmos. That's rational as in ratio, the heart of the Pythagorean cosmos. "Logos" means "word," the kind of reasoned word a Greek philosopher would speak. Christian historians deduce that the Gospel of John was aimed at Greeks when it turned the act of creation into an act of Logos: "In the beginning was the Word, and the Word was with God, and the Word was God." God creates the world through a sequence of words: "And God said 'Let there be light,' and there was light." Since God was still alone in the universe there was no one to hear him speak. What mattered was the divine power of words. Words are made out of breath, and breath is the essence of life.

The Greek and Christian emphasis upon the word may have culturally influenced twentieth century astronomers when they needed a name for the beginning of the universe. Humans rely on vision first of all the senses, and astronomers rely on light completely, so astronomers could have honored light by calling the birth of the universe "the Great Flash." Since there is no sound in space it is scientifically nonsensical to use an auditory term. Yet astronomers adopted an auditory image: the Big Bang. Was this a faint echo of the music of the spheres?

If a Big Bang falls in a cosmos with no one to hear it, does it make a sound? The cosmos was silent and the cosmos was deaf. Even after the cosmos started making sounds, the cosmos remained completely deaf. The sounds came from tiny pockets in the vast silence. As proto-stars accumulated gas, the gas began conducting sound waves. Even after the stars ignited they could still contain sound waves, but of course the stars could never contain ears. Planets and moons began making sounds, the sounds of Jupiter-dense atmospheres blowing and blowing, the sounds of asteroids crashing, the sounds of lightning and rain and volcanoes. But almost all of these sounds would remain forever unheard. Planets loaded with ear-like craters would remain stone-deaf. It was Annie Cannon and Henrietta Leavitt who represented the normal state of the universe, and it was the rest of the human race that was the extreme anomaly. Annie and Henrietta were hearing the true voice of the universe, the ancient and vast silence. The great silence had crept out of space and into their ears and taken over their heads. The great silence had taken on human form, a form that finally allowed the great silence to study itself and recognize itself and speak itself. The great silence became human in the way that Christ became flesh, to redeem an Earth that had forgotten its source. The Great Silence became the Word, and the Word was the sign language of the stars, and the Word was the Big Bang. The music of the spheres became voices with which to sing.

It was the nearly-silent ringing of a church bell that finally convinced Ludwig van Beethoven that he was going completely deaf. He had tried everything, but nothing had worked. He wrestled titanically against the Great Silence, but the Great Silence was far more powerful than he. The Great Silence wanted him and seized him. Beethoven fell into despair. He considered suicide. He moaned that he would be "forced to become a philosopher already in my twenty-eighth year."[36] A philosopher like Pythagoras? Beethoven withdrew into himself like a star collapsing into a black hole.

Of course! If Annie Cannon and Henrietta Leavitt and Harlow Shapley went to the symphony together, it must have been Beethoven. And not just any Beethoven, but the most completely deaf Beethoven, the Beethoven who composed music with enough loudness and drums and horns that even the deaf could feel it vibrating the floor and the chair and the teeth. It must have been the Ninth Symphony. It must have been the Beethoven who had gone beyond despair and suicide, the philosopher Beethoven, the Pythagorean Beethoven, the Beethoven who could hear in the great silence the music of the spheres. The Beethoven who could no longer hear the church bells but who still yearned to call out for celebration.

We can imagine Annie Cannon and Henrietta Leavitt and Harlow Shapley leaning forward as the Ninth Symphony winds and builds into its finale, as the singers and choir begin the *Ode to Joy*. This Joy is not just a feeling but a Greek goddess, the daughter of Elysium, the Greek heaven. Joy personifies a nurturing cosmos: "All creation drinks joy from the breasts of nature." Joy also seems to be Christian: "Brothers, above the starry vault, there surely dwells a loving Father." Never mind: the Romantics readily mixed pagan gods and Christian messages. The

Ode to Joy is a mystical vision beyond words, beyond religions, a vision of a joyous cosmos. Annie and Henrietta couldn't hear the words anyway. But perhaps they could feel the joy. They could certainly see the intensity of the players and the singers. Perhaps they saw the intensity of Beethoven composing music he could never hear. Or was it the Great Silence speaking through him? The Great Silence that shut out all other distractions so that humans could finally hear the music of the spheres. Perhaps as Annie and Henrietta gazed at the stars every day and every night they saw the notes of a majestic symphony. Was it the *Ode to Joy* they heard?

At the end Annie Cannon and Henrietta Leavitt and Harlow Shapley stood up. What is the sound of one cosmos clapping?

A Streetcar Named...

In the beginning, in the titanic struggle between chaos and order, in the sacred creation of human beings, light was almost always good. Perhaps in the cosmology of owls or bats light would be evil, but for humans light meant safety and warmth and cooked meat and crops growing tall and plump. The sun became a god, even in deserts, where the sun was a stern god. Humans built pyramids to the sun and built Stonehenges to testify to the sun's reliable track, back and forth every day, back and forth every year.

But for the mother of George Hale, light was evil, light meant chaos. She spent much of her time living in a dark room, the curtains drawn against the light, against the traffic and noise and people that troubled the light and gave her violent headaches. Mary Hale was frail and high-strung and chronically nervous, and it didn't take much to disturb her. Often the only stimulus she could tolerate was reading. She enjoyed reading to her son George, reading classics like Homer, reading about Apollo the sun god, Apollo the beautiful and true; and as she read, the lantern cast upon the walls the vague shadows of herself and George, of her moving hand and head and even lips, Plato's cave shadows that declared that what was seen might not be all that was real.

George's earliest memory was one of light as chaos. When he was three years old in 1871 a fire burned down central Chicago, including the house where he was born and his father's "fire-proof" office building. The flames shot monstrously high, like solar flares, broadcasting eerie shadows and smoke. George would compare this memory to light: "It was one of those curious flashes which strangely linger long after the throng of less tenacious illuminations have sunk into obscurity."[37]

Later George Hale discovered the idea that light meant order. Encouraged by his mother's reclusiveness and by his own frail health, which included fainting spells and violent headaches, Hale stayed indoors while other boys were out playing, stayed in his home laboratory playing with chemistry experiments and biology collections. At age fourteen Hale talked his father into buying him a telescope so he could observe a rare transit of Venus; Hale watched starstruck as the tiny dot of Venus moved across the sun exactly when and where astronomers had promised. He went on to observe a whole sky full of lights coming and going on schedule. Then he built a crude spectroscope out of a glass pendant taken from a candelabrum and fixed it to his telescope and pointed it at a candle flame. A rainbow erupted. A whole rainbow had been hiding inside plain white light. Hale built a more sophisticated spectroscope and pointed it at the sun and saw a rainbow with black lines upon it, the signature of various chemical elements inside the sun. The concept that mere light could carry such secrets far across space struck Hale like a revelation; he was "completely carried off my feet. From that moment my fate was sealed."[38]

Hale had been converted to an idea that was still new and struggling within the world of astronomy. Until now astronomy had consisted of mapping out the stars, their numbers and groupings and motions. But with the spectroscope you could discover the physics and chemistry of the stars, their contents and dynamics. No one imagined it yet, but the spectroscope would also enable astronomers to discover the distances and motions of galaxies, to discover the existence of an expanding universe.

Hale became obsessed with the sun and read everything about it he could find. In a book called *The Story of the Heavens* he was struck by a description of the vast scale of the cosmos:

> If a railway were laid round the sun, and if we were to start in an express train moving sixty miles an hour, we should have to travel night and day for five years without intermission before we had accomplished our journey.[39]

George Hale grew up during the late-nineteenth-century peak of human confidence in science, in not just the ability of science to understand the workings of nature but in the equation of science with human progress, the ability of humans to translate the laws of nature into mechanical inventions, like railroads. George's father owned an elevator company in Chicago, and after the Chicago fire, as a bold, new, steel-skyscraper Chicago emerged, it was Hale elevators lifting humans to new heights. Soon Hale elevators were lifting people up the Eiffel Tower. George would catch his father's ambition for achieving grand engineering feats, and he would devote his life to building ever-grander telescopes.

Yet science was questioning the universe in other ways. George was born nine years after Charles Darwin published his theory of evolution. George's mother came from a family of ministers and believed in a stern, Calvinistic god. She considered Darwin's theory to be an outrageous denial of the Bible, a denial that the universe had an intensely moral purpose, replacing it with an immoral chaos. As George read science and attended MIT and heard Alfred Russel Wallace speak, he struggled intensely with the contradiction between scientific evidence and his mother's faith; he went through doubt and guilt and depression.

George Hale's first major scientific invention had an inspiration as serendipitous as the apple falling onto Isaac Newton's head. Hale was pondering the trouble astronomers were having photographing solar prominences, which were thoroughly drowned out by the glare of the rest of the sun. Only during an eclipse when the face of the sun was covered could astronomers effectively study the prominences, but to catch an eclipse astronomers had to travel to remote places, and then they had only a few dozen seconds to make observations. Hale was pondering this problem one day as he was riding a Chicago streetcar and absently gazing out the window. Suddenly things were flashing at him. He was passing a long, white picket fence, and the objects behind it were flashing in and out of view. Hale started to wonder what a solar prominence would look like if you had an instrument with a system of thin slots for letting in the light of a prominence and filtering out all other light. Thus he invented the spectroheliograph, a major advance in solar astronomy; it made Hale famous at age twenty-one.

When Hale graduated from college soon afterward, no one expected his mother to attend, for she had become a complete recluse inside her darkened room. But she managed to come out into the light.

Light became Hale's lifelong obsession. He could never get enough light. Hale did important work as a solar astronomer, but his most important work was building the observatories other astronomers used to discover the expanding universe. Yet Hale was never satisfied with having built the world's largest telescope and he needed to start building an even larger one. His "divine discontent" to have more light would drive him into the same symptoms as his mother, and into several complete nervous breakdowns.

Hale started by building his own observatory in his own backyard, with a telescope with a twelve-inch lens. At about this time John D. Rockefeller gave some of his Gilded Age loot to found the University of Chicago; the university wanted an observatory and decided that the easiest way to get one was to absorb Hale's observatory. Hale saw the advantages of becoming the director of the observatory of a major university, and he had already recognized that the future of astronomy lay in big science—which needed big money.

In 1892 Hale jumped at the chance to obtain the world's largest lens, which had been ordered for a planned observatory in southern California and then been left sitting unpolished when a California real estate bubble burst and the observatory's rich patrons could no longer pay for it. Hale dreamed of building not just the world's largest telescope but the first observatory devoted to "the new astronomy," equipped with spectroscopes and sophisticated cameras and darkrooms and astrophysics labs. Such an observatory would open up the secrets inside the hieroglyphic light of the stars, open up depths of space never before seen or imagined, open up the extraordinary and beautiful order of the universe.

Yet such an observatory would cost a fortune, perhaps a third of a million dollars. The only way to fund it was to obtain donations from wealthy Chicago businessmen. Hale began making the rounds, only to find that businessmen saw no practical benefit in astronomy. Then someone suggested that such an observatory might appeal to the ego of Charles Yerkes, the most hated tycoon in Chicago and one of the most notorious robber barons in Gilded Age America. Yerkes had used every form of dirty dealing to build a streetcar empire, and then he used his power to assert political control over Chicago.

When Chicago burned down in 1871 Yerkes was a dirty-dealing banker and broker in his native Philadelphia. He was already involved in the streetcar business. Like many Gilded Age wheeler-dealers, Yerkes had built a financial house of cards full of speculation and debt and under-the-table deals. The thousands of houses burned down by the Chicago fire included Yerkes's house of cards. The destruction of a major business hub caused a national financial panic; insurance companies failed, banks called in loans, and investors sold off stocks to raise quick cash. Yerkes couldn't pay his debts, most seriously $300,000 illegally borrowed from the Philadelphia city treasury to make private profits for a city clerk. In trying to salvage his house of cards, Yerkes committed further improprieties. He ended up in prison.

When Theodore Dreiser was imagining a lead character for his novels about Gilded Age America, he decided that he couldn't invent anyone better than Charles Yerkes. Dreiser diligently studied Yerkes's life and translated it fairly accurately into his character Frank Cowperwood in *The Financier* and *The Titan*.

Toward the end of *The Financier* Dreiser has Cowperwood sitting in prison and looking up at the stars:

> He had never taken any interest in astronomy as a scientific study, but now The Pleiades, the belt of Orion, the Big Dipper and the North Star, to which one of its lines pointed, caught his attention, almost his fancy. He wondered why the stars of the belt of Orion came to assume the peculiar mathematical relation to each other which they held, as far as distance and arrangement were concerned, and whether that could possibly have any intellectual significance. The nebulous conglomeration of the suns in Pleiades suggested a soundless depth of space, and

he thought of the earth floating like a little ball in immeasurable reaches of ether. His own life appeared very trivial in view of these things, and he found himself asking whether it was all really of any significance or importance. He shook these moods off with ease, however, for the man was possessed of a sense of grandeur...Something kept telling him that whatever his present state he must yet grow to be a significant personage, one whose fame would be heralded to the world over...There was no more escaping the greatness that was inherent in him than there was for so many others the littleness that was in them.[40]

As a boy Theodore Dreiser loved roaming in nature and observing birds and animals and bugs and plants. His mother was a pagan who took the newborn Theodore outside for three nights to hold him in the light of the full moon and offer a chant for his future. His father was a fanatical Catholic for whom God was an avenging tyrant—even candy was sinful. At Mass his father told the child Theodore that God was present in the communion wafers, and Theodore cried out, "Give me God! Give me God!"

Yet Dreiser was born in the year that Charles Darwin published *The Descent of Man*, and ideas about evolution were thick in the air. At age twenty-three Dreiser read Herbert Spencer, and it "quite blew me, intellectually, to bits." Spencer was the prophet of evolution, preaching the new scientific concepts to a worldwide public. Spencer started with a thoroughgoing materialism, in which complex physical forces drove the cosmos, changing it into its current form. Spencer consolidated new discoveries from many disciplines—astronomy, geology, and biology—into an overall picture of a cosmos that was evolving, its stars and rocks and animals all loaded with change, all pointing to steady onward progress. Then Spencer added human society to this picture, and it too manifested the laws of progress. In human society as in the jungle, the force that drove progress was competition. It was Spencer who coined the term "the survival of the fittest" and largely started social Darwinism. In Spencer's take, the rich were the fit and the poor were the unfit who deserved their misery and perhaps elimination. Spencer was a perfect fit for Gilded Age America, and tycoons like Andrew Carnegie adopted him as their hero.

Theodore Dreiser wasn't so sure about social Darwinism. As a young man and as a newspaper reporter in Chicago he saw the misery of immigrants who, like his father, came to America hoping for a better life. Though Dreiser had largely given up on Catholicism, he remained disturbed by the loss of a moral universe and its replacement with heartless laws. Still, like most Americans, Dreiser was fascinated by the rise of new technological empires and by the men behind them. This tension between apprehension and admiration runs through Dreiser's novels about Charles Yerkes.

Dreiser begins *The Financier* with a young Yerkes/Cowperwood watching a struggle for survival in a local fish-market tank:

> He was forever pondering, pondering...for he could not figure out how this thing he had come into—this life—was organized. How did all these people get into the world? What were they doing here? Who started things, anyhow? His mother told him the story of Adam and Eve, but he didn't believe it...One day he saw a squid and a lobster put in the tank, and in connection with them was witness to a tragedy which stayed with him all his life and cleared things up considerably intellectually.[41]

Day by day the lobster devours the defenseless squid.

> "That's the way it has to be, I guess," he commented to himself. "That squid wasn't quick enough." He figured it out...It answered in a rough way that riddle which had been annoying him so much in the past: "How is life organized?" Things lived on each other—that was it.[42]

Cowperwood decides to go into banking and to be the top predator. Yet even as top predator Cowperwood has occasional intimations of human smallness:

> ...this Chicago fire. What a curious thing that was! If any one thing more than another made him doubt the existence of a kindly, overruling Providence, it was the unheralded storms out of clear skies—financial, social, anything you choose—that so often brought ruin and disaster to so many.[43]

Yet sometimes chaos could be very profitable. Using his political influence, Yerkes got an early release from prison, just in time for the financial panic of 1873. Like many of the panics and scandals of the Gilded Age, this one was all about railroads. The mighty Jay Cooke bank had sunk a fortune into building a northern railroad to the Pacific, only to suck itself into collapse. In the ensuing panic Yerkes sold stocks short and rebuilt his wealth. Then he moved to booming Chicago to start over.

Twenty years later when George Hale contacted Yerkes about the observatory, Yerkes was indeed the top predator in Chicago, loathed by businessmen, labor unions, politicians, clergy, and most riders of his monopoly streetcars. Yerkes had always craved social status but now he found that all his wealth and power couldn't buy dinner invitations in Chicago society. Yerkes immediately grasped that by having his name on the world's largest telescope he would win worldwide fame and force Chicago society to respect him.

Hale knew with whom he was dealing: he told Yerkes that the Yerkes Observatory telescope would be twenty-five percent more powerful than the then-largest telescope, and that it would be able to spot a quarter dollar at a distance of 300 miles. Yerkes said that he wouldn't give even a thousand dollars for the telescope unless it was indeed the world's largest. He wanted Yerkes Observatory built on a major street in Chicago so that everyone would see it and admire it. Hale was thinking of Mt. Wilson in California; he was among the first astronomers to realize that the desert Southwest was a far better location for telescopes than the smoky, humid, light-filled skies near the great eastern universities and cities. After some wrangling, Yerkes very reluctantly agreed to build the observatory at Lake Geneva, Wisconsin. This might be eighty miles from Chicago, but at least it was a resort area for Chicagoans and the site of the vacation mansions of Chicago's wealthiest. Hale also promised Yerkes that the telescope would be displayed prominently at the upcoming Chicago World's Fair of 1892. In *The Titan*, Theodore Dreiser portrayed Cowperwood's pride:

> The whole world would know him in a day. He paused, his enigmatic eyes revealing nothing of the splendid vision that danced before them. At last! At last![44]

When the Chicago newspapers announced Yerkes's plan to pay for the world's largest telescope, they teased that Yerkes must be searching for streetcar routes on Mars.

Before long Yerkes was begrudging the bills that Hale was sending him. Yerkes cared only about the status value of the world's largest telescope, not about nonsensical and expensive extras like spectroscopes and photography labs. Typically, Yerkes decided that Hale was a hustler and a swindler. Then came the financial panic of 1893; it too started when a railroad bubble burst, and it sent half the track mileage in America into bankruptcy, along with thousands of banks and businesses. Yerkes too was impacted and was even less willing to pay for the telescope. As further ego-bait, Hale hired a prominent architect to give Yerkes Observatory the grandeur of a Roman palace or cathedral. But another setback occurred when the newspapers reported that the president of the University of Chicago, which would own and operate the observatory, had attended a meeting of the Civic Federation that campaigned for clean government and fair business and that had long targeted Yerkes. Yerkes spitefully told Hale to go try to get his telescope money from such egghead do-gooders. When the observatory was nearing completion Yerkes feared that his enemies would try to sabotage it, and he insisted that it be guarded.

Finally Yerkes got his glory. For the dedication ceremony two trainloads of Chicago and scientific dignitaries went up to Yerkes Observatory, and when Charles Yerkes delivered a learned talk—undoubtedly written by Hale—about the progress of astronomy, the crowd roared their approval. A Chicago newspaper pronounced:

YERKES BREAKS INTO SOCIETY
Street Car Boss Uses a Telescope as a Key
to the Temple Door and It Fits Perfectly

From his first glimpse through the new telescope, Hale saw the order of the universe emerging. At first the only order he wanted to see was that of the telescope itself, proof that its optics were indeed the best ever. He focused on a planet, then a star, then a nebula, and again and again he saw a brightness and a crispness and details that he had never before seen in these objects. The telescope was a triumph of human engineering and human progress, which of course was the culmination of the progress of the universe.

Then Hale saw layer upon layer of cosmic order. He saw order from the atomic scale to the solar-system scale to the cosmic scale. He saw the order within light, the spectral signatures that were the same in light from near and far, proof that the same laws that ruled our own sun ruled the entire universe. He saw the order within the sun, the reliable patterns even in the midst of fiery chaos, the patterns of convection and rotation and electromagnetism, of granules and sunspots and prominences. He saw the order in the solar system, the steady clockwork cycling of the planets and moons. He saw the order of the stars, how they grouped into patterns small and large, from two stars circling one another to a hundred thousand stars in globular clusters to the billions of stars in our Milky Way galaxy. Hale didn't realize it yet, but he was also seeing the order of the galaxies and of the whole universe. Soon Yerkes Observatory was attracting bright students, and one of them, Edwin Hubble, would go on to discover the expanding universe and the Big Bang.

Edwin Hubble, whose grandfather was in the streetcar business in Springfield, Missouri, spent his later youth in a Chicago suburb, requiring him to ride Yerkes's streetcars to get into the city or to get home from his classes at the University of Chicago. For two years Hubble did research at Yerkes Observatory, placing him on the cutting edge of cosmology. He soon received a job offer from the new Mt. Wilson Observatory, George Hale's new world's-biggest telescope. It was a career opportunity that other young astronomers would die for. But Hubble refused it and joined the army as a private and went off to fight in World War One, a world of muddy, rat-infested trenches and no-man's-lands filled with lunar craters and rotting flesh. Here telescopes were used for spying on the movements of enemy armies; once Hubble was trapped in a swaying observation balloon while artillery shells flew by. Here the Big Bang was an artillery shell that exploded next to where Hubble was standing, knocking him to the ground, knocking him unconscious, giving him a concussion, leaving small cuts that became scars, leaving his right elbow permanently crooked, leaving him later moving the Mt. Wilson telescope with a crooked elbow, photographing the crooked galactic arms with a crooked arm, searching for the ultimate order of the universe with a fossilized chaos, homing in on the Big Bang with an echo of it that already knew its nature.

Theodore Dreiser once visited Mt. Wilson and met Edwin Hubble, whom he regarded, as he did all astronomers, as a priestly figure on a sacred quest. Dreiser looked through the Mt. Wilson telescope and was awed by the order of the universe, especially by the thought that the matter he saw through the telescope was the same matter he saw through the microscope, saw in cells and in the human brain and in himself. He rhapsodized that astronomy had proven the old religious faith in the oneness of humans and the cosmos.

George Hale spent World War One organizing a National Research Council to turn scientific research into weaponry. Hale considered psychology to be a science worth harnessing for war, and he hired Harvard psychology professor Robert Yerkes, a distant relative of Charles Yerkes. Robert Yerkes was establishing a laboratory for studying primate behavior, a lab that went on to become world famous. Yerkes studied the resemblances of primate and human behavior, including primate territoriality and aggression and status seeking.

To enlist British scientists Hale went to London, where he rode back and forth on the Underground, which had been built by Charles Yerkes looking for larger worlds to conquer, built through the same aggressive territorial machinations he'd used in Chicago.

Hale also hired Robert Goddard, who was soon shooting off rockets at the base of Mt. Wilson. Goddard dreamed of landing a rocket on

the moon. Before long, rockets were landing on London, and tens
of thousands of people were hiding in the Underground, feeling the
vibrations of the bangs above.

Shortly after Hale first looked through the Yerkes Observatory
telescope at the order of the universe, there was a big bang inside the
observatory dome. The many-ton hardwood floor, which could be raised
or lowered to allow different telescope angles, had fallen and shattered
itself; a support cable hadn't been properly attached. If Hale had been
standing on the floor when it fell he could have been killed. Hale was
also mortified by the possibility that the crash could have damaged the
telescope. He rushed to look through the telescope, and his heart fell
when he saw cracks on the lens. But it turned out he was seeing a spider
web at the tip of the telescope. A spider web that expressed the exquisite
symmetry of the universe. A spider web built by a spider moving in orbits
as perfect as those of the planets and stars. A spider web designed to suck
the blood and the life out of the universe.

Once I visited Yerkes Observatory, and a staff member kindly showed
me around. When he led me into the great dome I stood there looking
up in awe at the telescope. I was there with a girlfriend who spent her
days using an electron microscope to see the cosmos inside life, a cosmos
of molecules and cells; of exquisite symmetries; of orbits as perfect as
those of the planets and stars; of DNA spiraling like galaxies; of cells
proliferating in expanding bio-universes. She especially peered into the
cosmos inside human bodies, for she worked for a Nobel-Prize-rich
medical school, where they studied why the order inside life turned into
chaos, why reliable orbits jumped the track. There are even reliable orbits
in diseases, such as manic depression, from which my friend suffered.
Some years later I heard that she decided to commit suicide. To find a
new home for her cat she took out an ad in the newspaper, and the guy
who answered it took an interest in her too. They got married. She went
on living.

Sometimes George Hale became so severely depressed that he had
to cease work, resign his responsibilities, withdraw from people, seek
psychiatric help, and seclude himself in sanitariums or in quiet European
towns. He was tortured by headaches and nightmares. and legend says
he had continuing hallucinations in which an elf scolded him about
his life. Hale's turmoil included continuing questions about ultimate
realities, about life and death and afterlife, about his mother's religious
beliefs versus his own scientific training, about the public debates over
Darwinism and its moral implications. Theodore Dreiser became so
severely depressed, including spiritually depressed, that he once set out
to drown himself in a star-mirror river, but he was deflected by a chance

encounter with a jovial drunk. Charles Yerkes got depressed in prison, but it was the depression of shame and powerlessness, which would turn into a pathological need to dominate people.

When George Hale first arrived in Pasadena to build the Mt. Wilson Observatory he sought out Henry Huntington, the Los Angeles streetcar robber baron. They discussed building an electric railway to the top of Mt. Wilson. Yet for funding Hale relied mainly on robber baron Andrew Carnegie. Hale became so preoccupied with building the observatory that one day he was motorcycling down a Pasadena street and obliviously crashed into the rear of a streetcar, his motorcycle sliding under the car.

There was plenty of chaos in building the observatory. The mules that hauled supplies up the rugged mountain trail sometimes fell off and died, or sat down and smashed valuable instruments, or ate copies of *The Astrophysical Journal*. A workman too fell and died. The 1906 San Francisco earthquake nearly destroyed a telescope component being built there. Even when a dirt road was inscribed on the mountain, the task of hauling a giant but super-delicate telescope mirror up the steep twists and turns could induce a nervous breakdown in anyone. But Hale survived, and he went on to build an even greater observatory, Palomar.

The first time I drove the still-twisting road up Mt. Wilson, I'd turned on my TV that morning and seen a blue sky with a comet streaking across it, a comet breaking up into smaller comets. It was the space shuttle *Columbia* and its crew burning up. I delayed my drive up Mt. Wilson, and then I went with a vague sense that my drive was some kind of memorial to the astronauts. Through rugged ways to the stars.

When George Hale first gazed through the Mt. Wilson telescope he was horrified to see not Jupiter but several overlapping images of it. It seemed that the mirror was flawed. But it turned out that the workmen had left the observatory dome open during the day and the sun had shined on the mirror and heated it up. After many dreadful hours the mirror cooled off and the image cleared up.

Gazing through the telescope, Hale saw the order of the universe as no human ever had seen it before, planets and stars and nebulae and galaxies startlingly crisp and large and detailed. He saw the steady clockwork cycling of moons around planets and of planets around the sun and of stars around the galaxy. He saw the same exquisite order that emerged in the forms of life and that inspired Theodore Dreiser to marvel in *The Financier*:

> It would seem as though the physical substance of life...were shot through with some vast subtlety that loves order, that is order. The atoms of our so-called *being*...know where to go and what to

do. They represent an order, a wisdom, a willing that is not of us. They build orderly in spite of us.[45]

Dreiser mused thus in the middle of the trial of Frank Cowperwood.

As George Hale gazed deeper into the order of the universe than anyone ever had before, he gazed deep into the astronomical order, deep into the physical order, deep into the chemical order, deep into the geological order, deep into the biological order, deep into the sociological order, deep into the historical order, and there at the heart of it all he finally glimpsed the moral order of the universe. Or was this just another hallucination, like the scolding elf? He gazed deep into the moral order of the universe, the final reckoning of all order and all chaos, and there, moving like the clockwork stars—was it really what it seemed? Was the mirror distorted yet again? He squinted hard to make it out. There moving like the clockwork stars he thought he glimpsed, tracking through the void, nothing but—a streetcar?

Through the Looking Glass

For the longest time the universe could not see itself. To see itself the universe needed to have a face with which to see. The universe had plenty of faces, faces of rock with eye-socket craters staring blankly, sun-god faces without worshippers, gas-cloud faces swirling with dreams of solidity, faces of total darkness. Then there were faces of water, planets of oceans and lakes sparkling with sunlight and starlight, mirrors showing the universe to itself, mirrors no faces looked into.

After the waters of Earth had been mirroring the universe for ages, animal faces began appearing in those mirrors. Perhaps it was a primate face, not very far from becoming human, that first fully recognized its own reflection. Perhaps some animals had been skittish about looking into water because of the strangers staring back. But one face decided to test the face in the water and found that when she moved, it moved; when she smiled, it smiled; when she touched her eyes, it touched its eyes; when she reached out and touched the water, it reached out and touched her fingertip. She thought: *This is myself.*

She was even more correct than she knew. She was seeing water that had been flowing across the earth for ages, water that had flowed into the forms of life and into the form of her own body and face. Her finger's touch had set the water rippling, blurring her face but offering a clearer image of the ancient powers of flowing water: *This is myself.*

A million years or more later it was human faces staring into the ice of Ice Age glaciers and lakes, ice sparkling not only with sunlight and starlight but with campfires—the stars the life of Earth now carried in its hands. Even when humans saw in the ice only vague reflections they thought: *this is myself.* They were even more correct than they knew. They were seeing water that belonged to cycles much larger than the cycles of water: cycles of climate, cycles of continental drift, cycles of planetary motion. All of those cycles were tree-ringed somewhere deep within human bodies.

When humans started making spearheads they found that the best rock for sharp edges was obsidian, and they noticed that obsidian was so shiny that it offered reflections. Obsidian is a form of glass, usually dark black, cooked up by volcanoes. The Romans and the Olmecs polished obsidian to use as mirrors, though as Pliny the Elder complained, they were mirrors that turned people into shadows. The obsidian mirrors were more correct than people knew. Humans were seeing the volcanic fires within themselves, the geological powers compressed into their bones and eyes. And perhaps in their shadowy reflections they were seeing the truest portrait of their own mystery.

When humans started working metals they polished bronze into mirrors. The Chinese fashioned round mirrors because the moon, the sun, and the universe itself were round. Egyptian sculptors placed round mirrors on the head of Ra the sun god to signify the powers of light. Many peoples buried mirrors with the dead, perhaps as personal treasures but probably also as religious symbols. It was important for humans to reflect the order and light of the cosmos. The bronze mirrors were even more emblematic of the cosmos than people knew. Bronze is mostly copper, which is cooked inside stars. The copper inside human bodies was staring at the copper inside mirrors and seeing the long blacksmithing of the stars: *This is myself.*

With the perfecting of glass mirrors in the 1600s, mirrors went from being a luxury of the rich to being a household item for all. The most famous mirror manufacturer was the Saint-Gobain Company of France. Centuries later Saint-Gobain would provide the mirror for the Mt. Wilson Observatory telescope that Edwin Hubble would use to discover the expanding universe. But in their babyhood, Big Bang mirrors hung on bedroom walls to tell ladies who was the fairest of them all.

The spread of mirrors encouraged human self-examination, and not necessarily of the Socratic kind. Mirrors spread as jewelry, decoration, and architecture, culminating in the Hall of Mirrors at Versailles. As women and men alike became hypnotized by their appearance in mirrors, some Christians decided that mirrors so encouraged human vanity that they must be an instrument of the devil. Yet for a religion that often equated spirit with light, the symbolic possibilities of mirrors were irresistible. Dante has biblical characters hold mirrors to turn theological obscurity into the light of knowledge. Shrines to the Virgin Mary added mirrors to symbolize her purity. Johannes Gutenberg, before realizing how mirror-image type could create Bibles, made mirrors for pilgrims to wear on hats and gather the divine grace residing in cathedrals and relics. The power of mirrors to reveal truths of the soul lives on in monster movies in which a soulless Dracula leaves no reflection in mirrors.

It was Isaac Newton who first used mirrors in telescopes. Galileo's telescope used glass lenses to refract light, but Newton glimpsed the possibilities of reflecting light in concave mirrors. All of the great, universe-changing telescopes of the twentieth century would be reflectors. Newton built his first reflector with a metal mirror, but later telescope mirrors were made out of glass, polished to a precise curvature and coated with a reflecting material. At the end of his life Isaac Newton made a famous comment: "I do not know what I may appear to the world, but to myself I seem to have been only like a boy, playing on the seashore, and diverting myself, in now and then finding a smoother pebble or prettier shell than ordinary, whilst the great ocean of truth lay all undiscovered before me." Newton was more correct than he knew. Glass is made from sand, the same sand that makes up the seashore, the same sand that was once sandstone mountains and sandstone arches focusing the sun, sandstone that eroded into canyons and then dissolved and drained to the ocean and lay as sandy beaches, waiting to become new mountains. The seashore contains the same immensities of time as does the sky. It was immensities of time that became telescope mirrors and looked out into space and saw swirls of stars like the swirls of sand on beaches, like the swirling walls of sandstone canyons.

After all the time and energy humans had spent gazing into mirrors for the sake of human vanity, it was in quite a different spirit that they turned their mirrors out to space. It did not tend to serve human vanity. It was closer to the spirit in which humans placed mirrors in cathedrals, the spirit of seeking out infinities. Humans were still looking for themselves, but in a larger sense of self. Humans saw themselves in swirling carbon-rich gas clouds, in comets spraying out water like garden hoses, in stars cooking up the elements and exploding and fertilizing space, in distant solar systems, in galaxies that were mirror-images of our own, in the astro-footsteps leading back to the Big Bang.

The most important thing humans saw was that the universe now had a face.

The face of the universe, the star-freckled, galaxy-eyed, black face, noticed that someone was offering great mirrors to it. The face of the universe leaned forward and looked closely at Earth and looked into the mirrors.

First the face of the universe looked into the mirror of Isaac Newton's telescope. It was a small mirror but clear enough to show the largeness of space and the large population of the stars. The face of the universe stared at the stars in amazement and puzzlement, and a thought flickered: *Is this myself?*

Next the face of the universe looked into the mirror of William Herschel's 20-foot-long telescope. The face of the universe scanned along the Milky Way and saw that its fuzzy light was made up of stars, millions of stars. A thought flickered: *This is myself.*

Then the face of the universe looked into the 72-inch mirror of William Parson's telescope, 56-feet-long, called Leviathan. It saw nebulae everywhere, hundreds of nebulae, thousands of nebulae, and the nebulae seemed to be composed of stars. The universe was huge, and it was *myself.*

Next the face of the universe peered into the 100-inch mirror atop Mount Wilson, for now telescopes were being placed atop mountains to get closer to the stars. The face of the universe peered over the shoulder of Edwin Hubble and saw that the nebulae were galaxies as massive and star-packed as the Milky Way galaxy. There were millions of galaxies, and they were streaming outward from a common origin. *This is my fire-born and ever-growing self.*

Next the face of the universe gazed into the 200-inch mirror atop Mt. Palomar, and it was startled, for it saw gazing back at it something very different from space and stars. It saw two eyes, eyes filled with awe. It saw a human face gazing deep into the sky. It saw a human face spun out of space and stars and comets, and a thought flickered like a lightning bolt: *This is myself.* The face of the universe was a human face.

The face of the universe gazed into the Hubble Space Telescope and saw a pool of water. Reflected in that pool was a primate face, a face not very far from becoming human, a face testing its own reflection, a face fully recognizing itself for the first of many times. That pool of water has now become the mirror of the Hubble Space Telescope. Through this mirror the face of the universe gazed into the deepest space and time and evolution. It saw the star-grained beaches on which the prettier shells of life had washed up. It saw the gas clouds swirling, the stars blooming, the elements enriching, the planets sprouting life. It saw the galaxies flying away from the Big Bang, from a flash of light in which—even there—humans now recognized themselves. In the Big Bang humans saw a gleam in the universe's eye; humans saw the universe winking at itself.

The Homing Instinct

As dawn rolls onto the Atlantic coast like the tide, life is lifted from its sleep. Plants raise their leaves to soak up the sunlight, and animals raise their heads and then their bodies and move into the day.

Among the animals awakening this morning are two pigeons. They awake in their nest in the deepest corner of a cave. As daylight flows through the mouth of the cave, they stir and flex their wings and groom their feathers. They gaze at the cave, at the dust motes and insects floating in the bright air, at the patterns of light and shadow on the walls. They listen to the callings of birds outside.

Pigeons have been awakening to scenes like this for ages, for they prefer nesting on the ledges and crannies of cliffs—or in caves if available. Ages of pigeon experience assures these two pigeons that they are safe in a place like this. Yet this ancient instinct is now finding novel outlets. Recently in pigeon history weird new kinds of cliffs and caves have sprung up everywhere. This particular cave is unlike any cave that any pigeons have nested in before. Instead of rough rock and dirt, the walls are made of smooth metal plates, with wires flowing about. This cave is set not in a cliff, but is a giant box held in the air atop stilts. From one end open to the sky, the box tapers back to a cozy point. Strangest of all, the whole box occasionally rotates completely around. Yet pigeon instinct insists that this is a good dwelling place, so the pigeons are happy here. It is home.

The pigeons waddled to the mouth of the cave and sat there to sun themselves. They gazed at the world outside. After awhile they saw a car drive up, as it does almost every morning. A human got out. He stood there a minute, gazing toward the pigeons. Another car drove up, and another human got out. The humans made sounds back and forth. They gazed at the pigeons. Then they returned to one car and unloaded some stuff. The pigeons found this rather boring. These two humans came here all the time and were always loading and unloading stuff. The pigeons turned and waddled back into the cave, their feet scratching on the metal, their coos echoing from the walls.

Suddenly the pigeons saw the two humans climbing into their cave. The humans crept forward, nets in hand. The pigeons scurried about, trying to reach the exit, but the nets swooped down and tangled them wing and foot. The humans carried them outside and placed them into a cardboard box.

The pigeons sat fretfully in the cramped, dark box, listening to the voices of the humans outside. After awhile the pigeons felt themselves being lifted and carried and set down. They heard a loud slam. A steady rumbling sound started up all around them. The box began vibrating. They heard a frequent, heavy swoosh, swoosh, swoosh. Unfamiliar scents flowed into the box. The noises and vibrations went on for a long time, and when they ceased, the pigeons felt themselves being lifted again. The lid of the box flew open. The pigeons flew out. They flew a safe distance away from the unfamiliar man holding the cardboard box, and settled onto the grass. Looking about for the safety of their cave, they saw nothing familiar. They didn't know where they were.

Back at the microwave antenna, the two humans, whom other humans called Robert W. Wilson and Arno Penzias, went about measuring the intensity of the static in the antenna. Now that the pigeons were gone, now that Wilson and Penzias had climbed inside the antenna with buckets and mops and cleaned out the pigeon poop they suspected could carry an electrical charge, they hoped the static would be gone too. Though it seemed unlikely that the pigeons and pigeon poop had been creating that much static, Wilson and Penzias had tried everything else they could think of to account for it.

Yet there was no reason to worry. The pigeons only needed to tune into the energies flowing through space to find their way home. Right now their empty stomachs made them more interested in finding food than in finding home. They flew to a grain field, and like shoppers in a supermarket they wandered up and down the aisles, occasionally picking something out.

When their stomachs were full and they had rested awhile, the pigeons began looking at the sky. They rolled their heads about, trying to find the sun, but the sun was only a white blush behind heavy clouds. The pigeons switched devices. With the biological compass inside their heads, they focused on Earth's magnetic field, feeling its intensity and direction. Intuitively calculating the general direction of home, the pigeons rose into the sky, circled to check their bearings, and flew off.

This was more frustrating than ever. Penzias and Wilson had checked and rechecked the reading on the antenna, but it was no good. Getting rid of the pigeons and pigeon poop had resulted in a small decrease in the static, but most of it remained. It still read about 3 degrees Kelvin. Any new and complicated electronic device was bound to have some quirks to iron out, but Penzias and Wilson had talked with the engineers and fiddled with the equipment more than enough. They had gone over everything. The signal couldn't be coming from an external source either, like a satellite or radio tower. No matter which direction of the sky the antenna was pointed, day or night, the static remained the same. It was almost as if the heat was spread evenly throughout the sky. But that didn't make any sense. What could produce something like that?

The pigeons spent the night on a second-story window ledge in a small town. For awhile the sky cleared and they could look at the stars and compare them with the star map inside their heads and see exactly which way was home. When they awoke they went looking for food, and they discovered Mrs. Sara McPherson, who had been feeding bread to the pigeons in the town square for over twenty years. Eighty-eight years old, Mrs. McPherson had no close friends left in town. Her sister in California kept urging her to move there. But Sara had lived here her entire life and felt she didn't belong anywhere else. This was home.

The pigeons mingled with the flock on the town square for awhile. Then they tasted the energies flowing through space, lifted off, and headed for home.

They flew over the roofs of the town, over the roads and cars, the yards and playgrounds, the parking lots and cemeteries, into the countryside. They flew over farmhouses and barns, ponds and fields of crops, dozing cattle and busy tractors. They flew over woods and rivers, hills and cliffs. They flew over villages and cities, seeing amid the quiltwork landscape of nature the quilts of metal and cement. Over factories and highways their breath dredged in the fumes of things burning. Over farms and woods they breathed the scents of things growing. They saw the wind that buoyed their wings also rippling flags, laundry, and fields of wheat. Their wings chopped and chopped, carrying them to the one tiny place in a vast world where they could feel they were supposed to be.

Through his window the astrophysicist could see the evening light fading into darkness over the Princeton campus, and the multiplication of human lights. Yet he didn't take much notice of it, for he was gazing through another kind of window at a much vaster sky, at a much more dramatic transition between darkness and light. This window was simply a pad of paper in his lap. Every so often he looked back down at the pad of paper, studied the equations on it, and added another series of equations. As the numbers led him onward, he saw more and more clearly the first moments of the universe. He saw space ballooning outward. He saw the churning of powerful forces. He saw a menagerie of elementary particles. He saw the particles swarming about and fusing. And he saw energy of fantastic intensity.

The theory that the universe began with such explosive intensity was still new and unproven. But this astrophysicist had an idea. By knowing which particles had formed in the Big Bang, and by knowing how much energy was needed to form them, he could calculate the intensity of the energy of the Big Bang. Since the universe had been expanding ever since, this radiation should have cooled steadily as it spread out. If he could calculate how strong this radiation should be today, perhaps scientists could build instruments to measure it and thus prove that the universe really did begin in a Big Bang and was expanding.

After several hours of work, he had an answer. The radiation that was trillions of degrees hot in the Big Bang should be only a few degrees Kelvin today. That was awfully cold. The astrophysicist gazed into the night, feeling that this energy just had to be out there. He wished there was an instrument sensitive enough to detect it.

Suddenly, with alarm, he heard a strange fluttering sound at the window. He saw something moving just outside. He leaned forward—and he saw that it was just two pigeons settling onto his window ledge. The pigeons saw him and were alarmed, and they fluttered again and vanished back into the night.

As the sun arose, the pigeons flew on. Though flying over unfamiliar land, though mere specks in an enormous sky, they plotted their course with keen precision. They were drawing upon not just their own experience, but on the experience of all life over billions of years.

Perhaps even the earliest cells had a need to discern their location and learned to sense gradients of warmth and cold. As life evolved and began moving over large distances, often in seasonal migrations, it developed many ways to find its way, such as following scents, ocean currents, wind currents, or the shape of the landscape. But the most reliable guidepost was the energy flowing through space, the energy of the sun, the stars, and Earth's magnetic field. Life learned how to sense the magnetic field and to note the location of the sun and stars precisely, measuring them against an internal clock. Creatures of all sizes and shapes used these cosmic lighthouses to get to where they belonged.

The pigeons were but two specks of the massive tides of life flowing back and forth over the planet. These tides include turtles rowing hundreds of miles through the sea to lay their eggs on one special island, salmon leaping rapids to get upstream, herds of caribou fording dozens of icy rivers, clouds of butterflies flying thirty miles a day for a thousand miles, penguins climbing out of the ocean to find the exact pile of rocks where they had nested the year before, and arctic terns flying twenty-thousand miles from pole to pole. In spite of dangers like predators, exhaustion, lack of food, getting lost, physical barriers, no shelter, and deadly weather, the tides of life press irresistibly onward.

The pigeons began seeing a familiar landscape, and at last, in the midst of a clearing, they spotted their home and swooped down to it. They landed on the lip of the antenna and waddled inside. This was very odd—all of their poop was gone.

The pigeon's search was over. And a greater search was about to end here also.

The pigeons had no idea that their home was actually a giant version of the sensing device inside their heads. The antenna too detected the energies flowing through space, yet with far more sensitivity than any living organ could. This antenna was built for telecommunications, yet in its capabilities it was similar to many instruments humans had built for science, for surveying the universe, instruments that could measure with great precision the position of a distant star, the motion of another galaxy, or the energy output of a quasar. And though no one knew it yet, this antenna was uniquely able to measure the energy that remained from the Big Bang.

This antenna was the culmination of the four-billion-year-long process of life discovering its place in the world, the process of life catching on to new clues in the environment and developing ways to measure them more accurately. The sweeping back and forth of countless generations of creatures had woven this antenna on the loom of evolution. Out of that weaving, humans had arisen, and in their hands life progressed from making homing instruments out of flesh to making them out of metal, glass, and electronics. These instruments had to be much larger because humans wanted to discover their place in a much larger realm.

No one knows when life began using the position of the sun and stars, but in one way or another, life probably took up astronomy hundreds of millions of years ago. Yet these creatures only used the stars to find their place on a continent. Humans wanted to know their location in the whole universe. Beginning with notches carved on sticks, expanding to notches like Stonehenge, humans recorded the flow of lights in the sky, until now their computers were packed with measurements of billions of stars and galaxies. Beginning with vague twinges of wonder, humans spun stories about the universe, until now their stories were crafted out of intricate mathematics. Humans discovered they lived on a round planet going around a star, a star going around the galaxy, a galaxy that was one of a great many galaxies. And shortly, when humans realized that the 3 degrees Kelvin static in this antenna was the remnant of the Big Bang, they would know for certain that they lived in a universe expanding from a great explosion billions of years past.

Yet humans were searching for more than just their physical location in the universe. Other animals might be able to feel entirely at home in one particular spot on Earth, and such places were important to humans also. Yet humans required something more to satisfy their ache for belonging. Among animals which used the stars only as a map, humans appeared and began wondering what the stars were and where they had come from and what they meant for humans. Humans had been thrown into a universe of strange creatures and strange lights. Like pigeons dumped out of a box, they looked around and saw nothing familiar. They didn't know where they were. Humans longed to know where they had come from, and how they fit into the world. They longed to feel at home.

Humans have studied our physical place in the universe to discover our metaphysical place in the universe. We were taking the measure of the stars to take the measure of ourselves. We have drawn heavy philosophical inferences from what we took to be our physical location, such as being in the center of the universe, or being one speck among billions of specks in a vast darkness. Now, with this antenna, humans discovered our ultimate location, that we and all the universe were flying out of a Big Bang.

Perhaps not even this would be enough to satisfy the human ache to find where we belong in the universe. Perhaps in humans the need to belong has become unanswerable, for the location of the universe itself is unanswerable—the location of the universe in ultimate origin and meaning.

Yet for the life that began finding its location by dimly sensing differences of warmth and cold in the ancient sea, to have discovered our place amid all the warmth and cold, all the light and darkness, all the rushing bodies and emptiness of the universe, this was to have discovered a great deal.

The Big Bang Discovers Itself

Amid the ocean of energy flowing through space, there were a few particular waves of energy that one day met a strange fate.

The emptiness of space was all these waves of energy had ever known, except at the beginning of the universe. These waves of energy had been part of the intense energy of the Big Bang, and for the next 300,000 years or so they had swirled in a dense blob of swarming particles and energy, a blob swelling outward.

When the blob had expanded for 300,000 years, its energy became diffuse enough that the particles could combine into atoms. Gradually the atoms drew themselves into billions of galaxies. But while matter was concentrating itself and leaving most of space empty of matter, the energy released in the Big Bang continued flowing uniformly through space. Amid this energy were the particular waves of energy we are following.

These waves of energy didn't notice that the atoms around them were drifting away and clustering. These waves of energy couldn't notice anything. The only flew blindly onward, through space that was growing emptier and emptier, colder and colder. They flew through this emptiness for billions of years, occasionally passing a galaxy.

Then, after these waves of energy had wandered for nearly fourteen billion years, a galaxy moved into their path. The energy swept into this strange place, a place so different from anything it had ever known. Now this energy flew amid billions of brilliantly shining spheres. The energy might have been able to understand these spheres, for their swarming particles and energy was somewhat similar to the Big Bang. But there was something the energy could not have understood. Around these brilliantly shining spheres there seemed to be some smaller, darker spheres.

For thousands of years the energy flew on and on among the crowds of stars, and now and then it passed near one of those odd spheres circling the stars. Then, while the energy was passing a yellowish star, the energy found one of those smaller spheres directly ahead.

The sphere loomed larger and larger as the energy approached. The sphere was blue and brown and green and white. When the energy was very close, it began running into molecules. The molecules grew thicker and thicker. Drifting through this layer of molecules were some huge white puffs.

As the energy was passing one of these huge white puffs, it suddenly passed something else, something very bizarre. Moving through the air was a chunk of solid matter. The energy had never before encountered solid matter. This chunk of matter was made of the same kind of particles with which the energy had swarmed in the Big Bang, but now the particles were layered and linked together and arranged in highly complex patterns. Some of these particles were swirling about, yet swirling very gently, flowing in patterns that built new patterns. Most incredible of all, this chunk of matter had two thin appendages that were flapping up and down, propelling the chunk of matter through the air!

The energy flew on. It raced toward the surface of the sphere—a sphere that seemed to be a massive chunk of solid matter. The energy headed straight for a patch of greenness, greenness that turned out to consist of millions of small green things attached to brown stalks. And—what was this? Amid the greenness stood some kind of structure, some kind of box, tapering to a point. The larger end of this box was open and pointed toward the sky. After fourteen billion years of journeying through empty space, the energy fell into this box.

Attached to this box was a smaller box, inside of which were two chunks of solid matter. Like the chunk of matter that flew through the air, these two chunks consisted of particles layered and linked together and arranged in highly complex patterns, and many of these particles swirled gently in ways that built new patterns. But these chunks of matter didn't have wings for flying through the air. Instead, each had two long legs for pushing themselves across the ground. They also each had two other limbs, at the end of which were some wiggly things. On top of each of these chunks of matter was a knob with a lot of holes in it. Through two holes air rushed in and out. Another two holes absorbed light. Another two holes absorbed sounds. And one hole occasionally opened and closed, while something inside wiggled up and down.

The energy had never imagined that there existed a place as strange as this. It hadn't realized that matter could behave in such bizarre ways. The only matter it had ever known was particles swarming madly in the Big Bang. But now the energy discovered that during its eons of wandering through space, those particles had transformed themselves. This energy had known these very particles eons ago. This energy and these particles had swirled together in the Big Bang. They had sprung into existence together and for 300,000 years flown in a rapidly expanding blob. Then this energy and these particles had gone their separate ways. Now this energy met these particles again, and discovered what they had become.

Inside the smaller box the two clumps of matter moved about. Air and light and sounds poured through the holes in their heads. They levered themselves over to an instrument that measured the energy coming into the large box, and they placed close to the instrument the holes that absorb light. Inside their skulls particles and energy swirled in the strangest form of all, swirled to form thoughts about what the instruments said.

The instruments said that over billions of years the chaos of the Big Bang had crafted itself into cosmos and into consciousness. Chaos had flowed into a pattern, a life form, that could look back and see all the patterns, all the steps and creativity, by which chaos had become cosmos. At first these life forms had recognized nothing about the universe. Everything was strange to them. Yet gradually they recognized that they lived on a planet going around a star going around a galaxy that was one of billions of galaxies flying outward from a common beginning. They recognized that they were made out of many layers of particles and forces, and when they calculated how these particles and forces would have behaved when the universe began, they envisioned a great, intense blob of particles and energy swelling outward. Particles that had been created in the Big Bang were now creating an image of it. Particles that had swarmed about with a fierce intensity were now flowing very gently inside brains, calculating just how fiercely they once had swarmed. Particles that had been swept along on tsunamis of energy now realized that this energy would have diffused through space and become quite weak, but that it was still out there.

As the two humans stared at the instruments that measured the energy flowing in from space, there occurred a great reunion. These humans had known this energy long ago. These humans had been quite different then, only particles swarming madly. They and this energy had been thoroughly mixed together. They and this energy had sprung into existence together and for 300,000 years flown in a rapidly expanding blob. Then they had gone their separate ways. Now they met again.

For billions of years after the Big Bang, the Big Bang was forgotten. The energy and particles that had performed the Big Bang rushed into space obliviously. The particles formed galaxies and stars and planets, yet during all this work the particles were oblivious of what they were doing. On at least one planet the particles became alive, yet for ages life was oblivious of the cosmic odyssey of which it was a part. But in one particular form of life, the particles came to understand all they had been through. Particles that had performed the Big Bang now searched for and found the energy that had been there with them. The Big Bang remembered itself.

Channeling the Big Bang

Reaching out to touch the microwave oven, I thought of the Sistine Chapel hand reaching out to Adam. I thought of the powers of creation. There was no better symbol of the creativity of nature than seeds, for seeds hide within a speck the universe's hard-earned secrets of building order. I touched the button and let there be light: the seeds within began unfolding order, though not this time their preferred order of a corn stalk. The seeds banged and banged. The Aztecs took popcorn seriously, ritually adorning statues of their gods, including Tlaloc the god of fertility, with headdresses and necklaces made of popcorn. Perhaps the Aztecs tasted not just popcorn but the powers of creation.

The banging slowed and the light went off. I opened up the bag and felt microwave energy cooling off, pouring heat and odors onto my face.

I sat down and turned on the TV and tuned it to one of the channels that shows the Big Bang. Wow! The energy! The roiling chaos! The heat—no, wait: that was heat from my popcorn. But the chaotic energy on the TV screen was unmistakably the Big Bang. It frantically flashed and flickered and swirled. I leaned forward to try to discern patterns in it. My eyes went dizzy from the chaos. My senses went happy from the popcorn—the salt, the butter, the nutrients.

The energy of the Big Bang is still flowing through space, and in fact it makes up the large majority of the radiation and heat in the universe, far more than that from all the stars. As the universe has expanded, this energy has cooled off, downshifting into longer wavelengths, until today this energy consists of microwaves. The discovery and ultra-accurate measuring of this cosmic microwave background radiation was one of the most important projects of twentieth-century astronomy, proving first the reality of the Big Bang, then the precise intensity of it. Yet before scientists with ultra-sensitive instruments discovered this radiation, people were already watching it every night. The average TV set picks up the remnant of the Big Bang and turns it into the static or snow that fuzzes the screen when a set isn't tuned to a broadcast or cable channel. TVs pick up stray radiation from many sources, so only about one percent of a snowy screen comes from the Big Bang.

I stared into the Big Bang blizzard. The electron gun inside a TV shoots out volleys thirty times a second, but with no input of human-made images, the gun fires randomly, fires black and white dots. The white dots fly across the screen like the blizzard of galaxies flying away from the Big Bang. Look!—there goes our Milky Way!

These flashes of energy are the actors in my TV show tonight, a story profounder than the human-made stories on other channels. When our daytime persona goes to sleep the unconscious breaks through in dreams, and this channel was like our cosmic unconscious breaking through, reminding us of our primordial origins and depths. The energy of creation has acted out billions of forms and roles. It has donned the Greek masks of suns and planets, of volcanoes and rivers, of trees and dinosaurs, and of humans in all their further disguises, disguises that let us forget our Big Bang identity. I switched channels to a human-made story, and even here a touch of snow came through: it was the Big Bang animating human motions and voices.

I switched back to the Big Bang Channel and listened to the Big Bang speaking. The white noise was a faint echo of the Big Bang; it was cosmic surf rolling onto shore after fourteen billion years at sea. My TV was like a seashell that allowed me to hear a deep and ancient force that all along was my own pulse.

Like the ocean surf, the white noise was soothing. Yet the Big Bang waves found it exciting to roll into my ears and brain. Almost all of the Big Bang energy flowing through space causes no excitement at all. A tiny portion of it falls onto planets but doesn't excite the stone or air. Even in the antennas built just to detect it, the Big Bang energy stirred only super-micro reactions. But when the Big Bang energy fell into the antenna of a human brain it caused a far greater excitement than it had in billions of years. It set off a new Big Bang, a re-creation of itself. It set off not just an image of itself but an emotional resonance, a burst of consciousness. I was a bell being rung by the Big Bang, a school bell calling for attendance, a fire bell calling for excitement, a church bell calling for celebration.

My salty popcorn was calling for water. I poured $H2O$ into myself, hydrogen that was born in the Big Bang but which had now become the celebrant of the creative power within seeds. And it was good.

Ripples

A lake is a mirror that shows humans our true face. A human-made mirror may show a human face with greater vividness and detail, yet this is not necessarily our truest face. A human-made mirror may deceive us into believing that all we are is a human face, human-made, sharply defined, no mystery involved. Yet a human face reflected on the surface of a lake pulses and blurs, pulses with the motion of the water, of the wind, of fish just below the surface, of leaves and insects landing on the surface—pulses with forces that are much greater than humans, natural forces that human faces really do contain. A human-made mirror may show us only our social identity, the appearance we wish to present to other humans, but a lake blurs this identity, denies it, and tells us that we have a deeper identity made of water, wind, and sunlight.

A lake at night may be the truest mirror of all. The human face almost disappears; the contours and shades and hairstyles that make us individuals almost disappear; and we become a universal human shape, a dark shape defined by the stars around it, stars that offer us an identity even deeper than the identity of water and wind.

Tonight the surface of the lake is rippling, and so are the stars upon it. These ripples are being made not by wind or fish, but by the pebbles and small stones I am tossing.

I toss another little stone, and I hear the plop of its entry into the water, and I see the energy of my moving hand appear in the water in the form of rings moving outward from that point of impact. The outermost ring is the highest, with each following ring somewhat fainter. As the rings spread across the star-filled water, each star they cross bobs and flickers, and this allows me to trace the motion of the rings. One edge of the rings reaches the shore where I am sitting and collapses and rebounds weakly, and the opposite edge of rings moves steadily into the lake until I can no longer follow it. Again I toss a stone, and again rings spread outward and a hundred stars bobble as if on the blast wave of a supernova.

As I toss another stone and watch the ripples crossing the stars, I realize that these aren't the only ripples on the surface of the water. The stars too are ripples. It is not entirely correct to say that I am seeing stars, either in the sky or on the water, for of course the actual stars are far away in space, and what I really see is the light fired from those stars, light that hasn't been part of those stars for a long time. The light on the water is a tiny part of spheres of light that spread outward from the stars in all directions, like ripples spreading upon the lake.

Yet the violence of a star is much greater than that of a stone striking water, so the luminescent ripples from stars continue being produced for billions of years. When first launched from a star these ripples are more like tsunamis, possessing enormous energy. Yet the lake through which starlight travels is so vast, spreading the energy of starlight thinner and thinner, that by the time those ripples arrive on Earth they are far fainter, far less energetic, than the ripples I am making.

I toss another stone and watch my ripples mixing with the ripples made by stars. My ripples soon fade away, but the star ripples go on and on as steadily as they have for billions of years and will for billions more. The same stars have sparkled on Earth's lakes for eons, yet not on the same lakes; those lakes have come and gone constantly, come and gone by the millions. For a moment I feel like a terribly small and fleeting thing surrounded by giants. But as I watch the starlight in the water I recall that I have made more than just momentary ripples of water. I have also made ripples of starlight.

The atoms that now make up myself were once part of stars. I'll never know how many stars I include; perhaps only a few, or possibly hundreds that were scattered around the galaxy. My atoms raced about inside those stars for billions of years. They migrated back and forth through that turmoil, constantly colliding with other atoms and occasionally fusing together. Each of my atoms was built by a collision of lighter atoms, and with each collision there exploded from my atoms a burst of light. This light flowed in the rivers and tides of energy and matter inside stars, and then it reached the surface and flew into space.

Though this energy was never part of my human body, though I never used it to perform my actions and generate my thoughts, I still feel justified in saying that it was part of myself. If, of course, I define my "self" not in a human-made mirror, not just as a human body with a social identity, but as a deeper identity revealed by the pulsing surface of a lake, an identity that includes water and wind and starlight. If my face is blurred into the larger face of the nature from which we arose and which we still contain, then it is legitimate to say that "I" launched starlight into space. "I," after all, is simply the collective label of all the atoms in this body, the atoms that, long ago in a very different form, shed part of their substance as light. I can look upon that light with the same affinity I might feel for the whiskers or toenails I have shed. I can gaze into space and know that little pieces of myself are spread across the universe.

Yes, that light is still out there, somewhere, flying ever deeper into space. It has traveled across six or more billion light-years by now.

Some of the light shed by my atoms had a much shorter journey, for it was soon absorbed by clouds of gas and dust. Even now that light is locked in clouds drifting around the galaxy, or in the stars or planets those clouds became. Yet most of my light flew onward, passing star after star. Since my atoms were part of many stars and those stars were scattered through space, pieces of my light may have passed one another going in opposite directions. Each piece of my light was a small piece of a sphere of light expanding outward from a star, yet as these spheres expanded they merged into a common sphere expanding outward from the Milky Way galaxy. My light passed galaxy after galaxy.

Some of my light may have been caught by galaxies, like bugs flying into a spider's web. This light entered a galaxy and once again flew past star after star. Bits of it may have run into stars, been sucked into black holes, been absorbed by nebulae, helped to light up comets, or fallen onto the stone and dust of planets. Perhaps a few pieces of my light fell onto the surface of a lake. Perhaps my light fell onto a planet that was alive. If it landed on a leaf it might have been inducted into the services of life. It flowed through the traffic inside a plant, helping to run cells and expand leaves. When this plant was eaten, my light entered another creature and helped it live. My light was passed from creature to creature, joining many strange shapes and helping to generate many strange thoughts. And whose light now helped generate my thoughts?

Though it is infinitely unlikely, it is even possible that a piece of my light fell into a telescope or another kind of scientific instrument pointed out to space. My light may have became a speck on a photographic plate, which was handled reverently by strange fingers and scrutinized and discussed. Or my light may have raised slightly higher a number in a mass of data, or one point on a graph. A little piece of myself may have taught a distant species about the Milky Way galaxy.

But most of my light never entered galaxies. It only flew on and on through space, forming a sphere many billions of light-years wide and ever expanding. I will never know exactly where my light is and what it encounters, but I can know that some of it will fly onward forever.

While my light has been making its journey, my atoms too have made a great journey.

Partly this was a journey in space. When their stars were exhausted, my atoms were tossed back into space. They wandered past other stars, and occasionally rode the blast waves of supernovae. They drifted into gas clouds and back out again, into spiral arms and back out again. They flowed around and around the galaxy. Slowly they were converging, and finally they converged into one nebula and swirled inward and went into orbit around the embryonic star in the nebula's center.

The greater journey made by my atoms was a journey in form. While most of my light is no different than it was six or more billion years ago, my atoms have been transformed again and again. After being part of stars and then being solitary atoms adrift in space, they became part of a planet. Here they became stone and lava and sand, rain and rivers and lakes. When life began, my atoms began joining much more intricate forms, joining DNA and leaves and hearts and eyes. My atoms passed from body to body, mission to mission, helping plants catch sunlight, helping birds fly, helping turtles break out of eggs and crawl into the sea, helping dinosaurs roar. And finally all my atoms came together and formed me. In the form of a human brain, where ripples of electrical impulses break out and flow through billions of neurons and fade away, atoms that once made ripples of light in space are now making ripples of thought.

I toss another stone into the lake, and as it strikes the water and starts ripples flowing outward it also plunges into another lake, an electrical lake inside my head. Ripples of electricity flow through my neurons, creating an image of the surface of the lake gently fluttering, the stars upon it glimmering. Then the ripples of water fade away, and so do the corresponding ripples in my head.

Yet as these electrical ripples fade, some new ones begin to flow. These ripples flow not in response to a stone plunging into water, but in response to the action of some stars billions of years ago. These ripples create an image of those stars, and of waves of light spreading into space, flowing on and on.

Billions of years after my atoms generated that light, they now are able to generate consciousness of what they once were and what they once made. After a long darkness, my atoms are generating light again.

I learned forward and gazed down into the water. The stars upon it shined so brilliantly that I had to squint. In this brilliant light I saw my own face more clearly than in any human-made mirror. My face pulsed with light. My face pulsed with the heartbeat of stars.

Celebrating the Big Bang

Every year in the town where I live we celebrate the Big Bang. We set off huge re-enactments of it in the sky, bursts of light that swell outward. To watch the Big Bang, many people gather on the edge of the mesa that holds the telescope that helped to discover the Big Bang. Every town should have a telescope that discovered the Big Bang.

This telescope is at Lowell Observatory in Flagstaff, Arizona. It was built in the 1890s by the greatest telescope maker of the time, Alvan Clark and Sons. It was one of the last great refracting telescopes ever built, as the age of glass lenses gave way to the age of mirrors. The Lowell telescope has a 24-inch lens, and because of its location in the high Southwest, it often outperformed the world's largest refractor, a 40-inch at Yerkes Observatory in muggy Wisconsin. Today the Lowell telescope is an antique, no longer used for science. But this means that it is open to the public. Professional astronomers have to wait years for one night at other telescopes famous for their discoveries, and even then they don't get to look through it with their own eyes. Here, any night they wish, anyone can put their eyes to one of the most famous telescopes in the history of astronomy. It was, to begin with, the telescope through which Percival Lowell saw canals and a civilization on Mars, filling the popular imagination with Martians for half a century to come. More importantly, it was the telescope with which Vesto Slipher discovered the redshifts of the galaxies, redshifts that Edwin Hubble used to prove that the universe is expanding.

In the museum building next to the observatory is the spectrograph Slipher attached to the 24-inch telescope to measure the spectra of galaxies. A sign explains how this spectrograph came to be here. Percival Lowell had borrowed a telescope from telescope maker John Brashear and thought he should clean it before sending it back. Lowell sent a boy to buy some cleaning alcohol, but the boy bought the wrong kind of alcohol, which damaged the surface of the telescope lens. Lowell felt so badly about this that he tried to make up for it by commissioning Brashear to build a state-of-the-art spectrograph. By such accidents, universes are changed. Vesto Slipher now had the world's best spectrograph, and he used it to make the first systematic spectrographic study of galaxies—except that Slipher didn't know he was studying galaxies.

Percival Lowell thought he was setting Slipher to work studying the formation of solar systems. Lowell thought that spiral nebulae were nearby solar systems in various stages of evolution. Lowell wanted to prove his theories about Mars, about how our solar system had evolved to leave Mars older than Earth and drying out and in need of canals. But other astronomers had wondered if these nebulae might be entire galaxies far outside our own Milky Way galaxy. As Slipher measured spectra, whose shifts toward the blue end or red end of the spectrum revealed motions toward or away from us, he was startled to find motions everywhere, fantastically fast motions, mostly motions away from us. Slipher was seeing the expanding universe, though he wasn't quite ready to recognize it.

When Albert Einstein developed his General Theory of Relativity, he was startled to find that his equations implied that the universe must be expanding. Einstein was so incredulous at such an idea that he invented an artificial equation to cancel out his own math and leave the universe motionless and stable. Like Einstein, Slipher and many other astronomers had a hard time coming to terms with the idea of an expanding universe, even as they were finding the evidence for it. In the absence of proof that spiral nebulae were other galaxies, and in the absence of accurate measurements of how far away the nebulae were, there was no way to correlate Slipher's redshifts with the distances of nebulae, no way to map out an expanding universe. There remained a possibility that the redshifts represented flawed theory or flawed observing techniques. Vesto Slipher's personal situation made him especially cautious. He was an observer, not a theorist. The first spiral nebula he observed was Andromeda, which showed a blueshift, contradicting the redshift pattern he would soon begin to find. Slipher was acutely aware of the damage that Percival Lowell's fantasies about Martians had done to the reputation of Lowell Observatory, and Slipher was reluctant to attach the name of Lowell Observatory to an even more fantastic scheme. Slipher's reticence meant that he would never get the credit he deserved for the discovery of the expanding universe.

It was Edwin Hubble who became famous for discovering the expanding universe. Yet Hubble's discovery was, first of all, his discovering that the work of two other men, Harlow Shapley and Vesto Slipher, could be combined and developed into a bold new map of the universe. Hubble used Shapley's Cepheid variable method to prove the reality and distances of other galaxies, and then he used Slipher's spectrographic measurements to correlate the distances of galaxies with their motions. Hubble found that the farther away a galaxy was, the faster it was moving—and moving away.

For anyone, looking through any telescope should mean looking back in time. For me, looking through this telescope is even more associated with seeing eons. I first looked through this telescope on the night of the day I first saw the Grand Canyon. All day long I had been staring into the past, the mile-deep past, the cosmological eras imprinted in rock. The black rock at the bottom of the canyon is 38% as old as Earth, 13% as old as the universe itself. It belongs to the same era as a galaxy 1.7 billion light-years away. Above the black rock, more recent eras reached toward me, eras in which ranges of mountains and ranges of life had appeared and disappeared. I had been staring into a huge empty space that was steadily growing larger. That night at the observatory my boots were still dusty with Permian time and my mind was still filled with layers of time reaching deeper and deeper. I was still seeing the black foundation of my own life. When I put my eye to the telescope, this was the blackness I saw, a blackness rich with time and power and creativity and beauty. Every star glowed with Grand Canyon sunset beauty. Ever since, things I've seen through this telescope have rung with a Grand Canyon echo.

Years later I moved to Flagstaff, and I enjoyed the privilege of visiting Lowell Observatory as often as I liked. Knowing its history, the 24-inch telescope tinted everything with expansion. Even the redness of Mars or a red giant star reminded me of the redshifted universe. The spectrograph was my own mind. I'd look at a galaxy that had poured its light into Slipher's spectrograph and I let its light pour into me, into billions of cell lenses and into blood that constantly expanded red and contracted blue as if the universe's whole history, its birth from its mysterious heart and its expansion and depletion and possible contraction, was constantly being proclaimed in me, which indeed it was, for a red blood cell rushing through a brain contains the whole momentum of cosmic evolution, all its ages and toil and genius. In my spectrographic brain the galactic light prismed forth the red glow of memories much older than my own yet in which I could remember my own past, my astronomical past.

As time passed, I visited the telescope less frequently. One of the curses of human psychology is that we quickly habituate to things, even the most amazing things. I took the telescope for granted, just as I take it for granted that I live at the one moment in human history when we are discovering the cosmos, just as I take it for granted that the universe is here at all and that it gave rise to life, and that one of those lives is me. But now and then I was reminded that I lived in a town on the edge of forever.

I'd be standing in line waiting for a look through the telescope, when a child stepped up for a first-ever look through a telescope. Maybe he was a vacationing city kid who'd seldom seen a clear sky, or maybe she was a Navajo child who lived on a remote desert road and who barely understood the purpose of a telescope. The mere sight of the telescope—

thirty-two-feet long and seven tons, painted silver—was enough to excite them. With one look into the telescope, the children exclaimed their surprise and delight and wonder. They didn't need to know any theories of the universe or histories of astronomy to be enthusiastic. They helped to rekindle my own appreciation.

Not everyone came here planning to look into the sky. In downtown Flagstaff the turnoff to Lowell Observatory on Mars Hill is less than a block before the turnoff to the Grand Canyon, and sometimes people heading for the canyon turn too soon and discover they are heading up Mars Hill. Some people turn around, but some continue up the hill and gaze into the ancient strata of the sky.

As people gaze through the telescope at the full moon, the moonlight casts a circle of light onto their faces, leaving an eye glowing in the dark. The telescope is a microscope through which the moon stares at us. When it is my turn to look, I feel the stare of the moon, the planets, the stars, the galaxies, all of space and time. With an instrument powerful enough to probe distant galaxies and glimpse the structure of the universe, the universe probes me. What does the universe see? Does it see something incredibly small and insignificant, the way everything looks tiny when you look through binoculars the wrong way? Is the universe's vast emptiness funneling into me, voiding me? Or does the universe discover, amid the vast blindness and faces of rock, a bright living eye; an eye that is the universe's own; an eye that through the telescope glimpses the cosmos it has always contained; an eye with a power greater than any telescope or gravitational lens, the power to focus the whole spreading cosmos into a child's cry of affirmation.

Every year in Flagstaff we celebrate the Big Bang and the expanding universe. We celebrate it on the fourth day of July. One 4th of July I drove up Mars Hill, passing a crowd of people on the mesa rim. I went into the observatory dome and looked at Jupiter, which in a couple of weeks was going to be struck by a comet co-discovered by a Flagstaff astronomer, Carolyn Shoemaker. Then I stepped outside and gazed at the horizon. I waited, as space and time once waited. Suddenly the black sky was shattered by a point of brilliant light. It was the Big Bang. From a point, the light flew outward, expanding into a huge ball of glowing red spots. Each glowing spot was a galaxy, redshifted, packed with stars and planets and emerging life. When this Big Bang faded, another was set off, then another. The bangs were several miles away, downwind, so they were soundless, at least for a few moments. I wandered back and forth to find a viewpoint unobstructed by pine branches, and I ended up sitting on the railing of Percival Lowell's tomb. An observatory staff member joined me and said that if the dome wasn't being used, the staff would climb up on the dome to watch the fireworks. I imagined Vesto Slipher climbing onto the dome beneath thousands of stars and watching the universe expand.

For another half hour the Big Bang ignited and bloomed, in many colors. The rainbow Big Bangs ignited in me a feeling I wished I felt all the time. I wished this feeling was always ignited by the lights in the telescope, all of which, including Mars and the moon, were flying fragments of the Big Bang. I felt enormously privileged. Partly this was the privilege of place, of being able, anytime my mind was cluttered with worries and superficialities, to plug myself into the Big Bang and be x-rayed by the universe's ultimate questions. Partly this was the privilege of time, of being present when billions of years of forgotten events are being remembered. Mostly it was the privilege of being alive. In the presence of both the Big Bang and the place of its discovery, I felt the enormity and improbability of the journey of raw energy into planets and pine trees and people. I gazed back through eons at my own birth. If only every moment of living contained the intensity of the Big Bang. If only the universe hadn't forgotten that intensity of being. If only I could remember it, every moment would glow with mystery and celebration.

Helios

The ancient Greeks knew that if they wanted the sun to rise, they had to give thanks.

For the Greeks the sun was the god Helios, who every day rose from his golden palace in the east and drove his winged horses and chariot across the sky. Beyond the western sunset horizon, Helios boarded a golden boat and sailed the river Oceanus back to his palace in the east. Helios was daily repeating his role in creating life, for the earth had been nothing but bubbling mud until Helios shined upon it and the earth sprouted plants and animals.

To honor Helios the Greeks built the Colossus of Rhodes, one of the seven wonders of the world, a 100-foot-tall statue of Helios, supported by iron beams, that presided over the harbor of Rhodes. The Colossus of Rhodes would help inspire the Statue of Liberty over two thousand years later, and would be mentioned in the Emma Lazarus poem, "The New Colossus," upon Liberty's pedestal. The Greeks also built temples to Helios, where they sacrificed white horses and white rams. The Greeks had to please and appease Helios because, like most Greek gods, he was far too human, subject to temptations and jealousies and anger. In *The Odyssey* Helios wrathfully killed the crew of Odysseus when they ate some of Helios's cattle. In 226 B.C. an earthquake toppled the Colossus of Rhodes, and the oracle of Delphi said it was because Rhodes had angered Helios. The Greek universe, like most pagan universes, was a very unstable place.

To escape from such chaos humans often readily traded pagan universes for monotheism, which offered greater benevolence, and for the scientific worldview, which offered greater stability and control. Yet there is something to be said for a universe where humans have to offer daily thanks for sunlight in order to keep the sun going. It is too easy for humans to take the sunlight, and life itself, for granted.

In 1868 astronomers used a spectroscope to examine the chemical composition of the sun, and they discovered an element that had never been found on Earth. They named it for Helios: helium. Twentieth-century astronomers realized that it was the fusing of hydrogen into helium that generated sunlight and thus life on Earth.

Astronomers also realized that helium, like Helios, played a key role in the creation of the universe. While helium could be generated inside stars, most helium was created in the Big Bang. The Big Bang first created hydrogen, then began fusing hydrogen into helium. As the Big Bang expanded it lost the energy required to create heavier elements. The universe was born as a hydrogen and helium universe. It would take stars and eons to turn helium into carbon and oxygen, the elements of life.

In honoring the namesake of helium the Greeks were more correct than they knew. It's even more appropriate for us to honor helium and the stars and the Big Bang through which helium helped give us light and life. Because humans have discovered the Big Bang cosmos rather quickly, human culture has been a bit slow to adjust to it and to devise ways of celebrating it. I would like to propose a great public ceremony for celebrating the Big Bang cosmos. This ceremony will be full of helium.

Helium was eventually discovered on Earth, but it's a rare element. Since helium is much lighter than air, free helium likes to escape into space. But helium gets trapped inside the earth, and humans learned to extract it from natural gas. I suggest that we take a lot of helium and watch it trying to return to space, as if it was longing to rejoin the cosmos from which it came.

More specifically, we'll get some balloons and fill them with helium. Really large balloons. The balloons will expand just like the expanding universe. The helium will be reprising its Big Bang role. The balloons will have the shapes of animals, including some of the animals sacrificed to Helios. The helium will take on animal shapes just as Big Bang helium evolved into animal shapes. The animals will rise just as life evolved out of the formless earth. A balloon will symbolize both the simple beginning of the cosmos and today's biological complexity. The balloons' eyes will see with the wisdom of fourteen billion years.

The balloons will fly through the air just like Helios, paraded through city streets crowded with a million people. Perhaps the parade should be held in New York City so that the New Colossus can preside over it. If the parade needs corporate sponsorship, perhaps one of New York's famous department stores would be interested. We could call this celebration "Thanksgiving."

We'd be giving thanks to helium, and to the sun, and to the Big Bang. We'd be giving thanks that the Big Bang gave rise to animals, and to us. We'd be carrying on in new form the ancient human need to remember and honor our origins, to rejoin the cosmos from which we came. It would be uplifting.

A Birthday Wish

The candles on my birthday cake flickered, sparking me to ponder time past, as birthdays tend to do. Yet when you live in astronomical time, flickering lights can invoke more than just candles, and time past can include more than just your human past. As I stared at the candle flames, I was hypnotized into seeing stars, stars long past, the stars from which I was born. I saw my original birthday candles.

I was seeing the stars in which my atoms had swarmed, the stars in which my atoms were formed. I was seeing hydrogen atoms tying themselves into larger atoms like ribbons on a birthday present. Hydrogen became carbon and oxygen, the elements of life. Carbon and oxygen became silicon, which began counting down the remaining months of a star's life until it died as a supernova, and which later on, as sand, would run through hourglasses to count the hours of human lives, and run through beaches to count the ages of a planet's life. Oxygen became sulfur coating a star, sulfur that later would coat the matches by which my birthday candles were lit. My atoms built up into iron, which would serve me loyally to carry blood from my heart, but which triggered a heart attack in stars, supernovas that blew out the stars and blew their elements into space.

I closed my eyes to blow out the candles, blow them out with carbon dioxide—carbon and oxygen. I thought of the many creation stories in which the breath of a god brought about the creation of the world. Often it was breath formed into words, into incantations with powerful magic. In Egyptian myth it was a single word, Ra's secret name, that gave Ra his divine powers and that enabled him to speak words that turned the primordial waters into land, animals, gods, and humans. In Mayan myth there was total silence until the gods Tepeu and Gucumatz spoke the word *Earth!* and the earth was created. The Judeo-Christian god turned the Word into the World; he needed only to say "Let there be light," and there was light.

But of course to make a birthday wish, we are supposed to blow out the lights. We are supposed to reenact the supernovae that blew out our stars and sent our elements flying into space and toward the future Earth and toward our birth. Yet in coming up with a birthday wish, I tried to give it some of the dignity of gods breathing spirit into matter. Birthday wishes seem to retain some old magic, for of course we are supposed to keep them secret, just as the gods' names and spells had to remain secret. I envisioned the ancient stars from which I was born and intoned:

"May you fill the darkness with light, with the power and beauty of light, and fill atoms too with that power and beauty. May you take the hydrogen from the Big Bang and tie it into carbon and oxygen. May you count off the elements like rosary beads; may you climb the cathedral stairway to life. But—may you not make light forever. May you blow yourselves out, every one of you. May you wish upon a star to not be stars anymore. May you become Earth and life. May you awaken one day and be amazed by what you have become and by the journey you have made. May you rekindle some of your old intensity with which to feel amazed at being alive—at being itself. May you learn to celebrate your lives. May you learn how to become a chocolate cake."

I blew forth a long, scanning breath. Then I opened my eyes. All of the candles were out. I gazed at my chocolate cake. It worked!

From each candle a thread of black smoke curled up, black carbon dispersing, a bit of which I breathed in to become myself.

There were fourteen candles, one for each billion years of the universe's fourteen billion years. Where does the time go? One day you are just a Big Bang, a tiny baby universe, and before you know it you are all grown up and looking back wistfully.

I looked back at the vanished stars from which I was born. I looked back at the journey I had made, at layer upon layer of order and creativity. I looked at layer upon layer of chocolate. I pulled out a slice of chocolate cake and tasted it. I tasted fourteen billion years. I tasted the work of vanished stars, stars which now kindled in me a bold new flavor of energy, the sacramental light of celebration.

Impression: Sunrise

The house was dark, darker than the night sky above it, whose darkness flickered with stars. Against this sky the house was visible only by its rectangular absence.

Then, in the upper corner of that darkness, a small flame appeared. It flared for an instant and then settled into a wavering spot of light. The flame moved downward and to the left, and from it arose a larger flame, which gleamed faintly from the glass and metal of the lamp that held it. The flame glowed onto the objects around it, vaguely suggesting their shapes, and it glowed again from the windowpane. The lamp rose into the air and drifted away, leaving the house dark for another minute, and then the flame drifted into a window on the first floor. The light settled in the lower center of the window, and from there it glowed out into the night.

Light. Instead of emptiness, instead of inertness, instead of a universe forever night, there was energy, there was action, there was form. Light is the revealer of the universe's secrets, of matter's inside and outside. Light is the energy that composes all matter, showing its real face. Such energy is the mover that makes flow all the events of the universe. And when that energy flows naked through the air, it also discloses the external forms of the matter it touches.

Light. The light from the lamp gently gathered the shapes in the room around it and carried them away, out the window and into the garden, where they mingled with shapes that had been transported across much greater distances by much greater light. Planets were floating here, and thousands of stars bright and colored. Whole galaxies had been carried here across millions of light-years of night.

Light. In the glow in the window a new shape appeared, the contours of a human face, looking out. Though the face remained mostly submerged in the dark, the light revealed a thick beard, a forehead on which many years had been written, a cheek, one side of a mouth, and an eye, an eye that sparkled faintly. Light. The eye gazed into the garden, gazed up at the sky, where it saw no clouds, only stars. His mouth curled into a smile. Light.

Later, when the stars were fading and the first glow of dawn was rising over the eastern horizon, a door opened and the man emerged. Over his shoulder was strapped a box, and under his arms he lugged a folded easel and a large canvas. He climbed down the stairs and walked between long rows of flowers and beneath wooden arches tangled with vines. He pushed through the gate and crossed the road and the railroad tracks and stopped beside the pond. He gazed at the water, still black, and at the lilies, still only vague shapes in the dark. He followed the path along the pond, walking slowly, pausing to consider, and continued up to and onto the bridge, pausing on it to gaze into the water. He followed the other bank back again, gazing across the water at the gathering light. At a point where fifty feet of water lay between him and the eastern sky, he stopped and set down his paint box and set up his easel and canvas.

By the ceaseless turning of the spheres, the darkness of the sky turned into dawn. The blackness became a dark blue, which lightened into a lighter blue. Other colors appeared, diffuse blushes of pink, yellow, and red, which flowed in size, shape, and brightness, flowed slowly at first and more rapidly as the sun approached. With the awakening sky emerged the colors and shapes of the land: the hundred shades and thousand shapes of green; the textured browns of tree trunks and branches; the yellow of nearby fields of wheat; the bright colors of the flowers around the lake and near the house. From the planet's dwindling shadow each object began to untangle its own shadow and spread it upon the ground or upon its neighboring objects, shadows vague now, but soon to grow distinct. The increasing light gave the pond a dim glow that would soon intensify into brilliance, and gave the pond not its own color but every color around it, predominantly the blue of the sky and the green of the trees and lilies.

As the light increased, Claude Monet took out his brushes and paints and squeezed blobs of color onto his palette. He stood and waited. He watched the flowing of light and color over the land and water. He heard the talking of birds, eager for sunrise. Then it came, a point of total light on the horizon, a sudden wave of light over the land. He touched his brush to the palette, lifted the brush toward the empty canvas, and pressed upon it one short stroke of bright yellow.

By the ceaseless turning of the spheres, the darkness of the sky turned into dawn. All three moons were visible, the nearest moon large and glowing an icy blue, the other two moons merely sharp lights. As the sky turned pink and then red, the moons faded. The land too showed some redness, though this was but a vague hue amid the grey. The plains of dust were grey, and the rocks scattered upon them were grey. The crater rims were grey, and the distant hills and mountains were grey. As the sun appeared, it spread a further greyness behind every rock and hill and crater rim. The increasing heat stirred the wind to sweep the dust into clouds and race them across the land, and this dust storm would grow steadily thicker through the morning and take back the brightness of the morning sky.

As the brush strokes added up into the hundreds, forms emerged out of the blankness. There was water scintillating with sunlight, shining gold and silver. There were green lilies floating on the water, their flowers white, yellow, pink, or crème, and they glowed again in reflection from the water. Monet was working quickly, hurrying his gaze from horizon to canvas, hurrying his brushes from palette to canvas, for he was seeking to capture a moment, an appearance that would soon change into something else. Instead of drawing sharp lines between objects and homogenizing the colors within each, he mixed dots and dashes of different bright colors, letting them fuse into luminous patches and bursts of sunlight. This impressionistic style was meant to portray not objects as much as the behavior of light, the way light made the world vibrant. It was meant to recreate the visual and emotional impact that a real scene stimulated in the human eye and brain.

By the ceaseless turning of the spheres, the darkness of the sky turned into dawn. Already the sunlight was creeping along the ring that surrounded the planet, setting its millions of chunks of ice glowing as if they were a single arch. It glowed softly, for the sun was far away. Sunlight also advanced across the planet's surface, which is to say across the outermost layer of the clouds that formed most of the planet's bulk. The clouds welled out of the planet's turbulent depths and formed hundreds of bands that swept across the planet and mixed as giant swirls and spirals, which the sunlight set glowing with dozens of colors.

Years ago, when impressionism was too new and strange to be accepted by the critics and public, Monet and thirty other explorers of the new style, among them Renoir, Degas, Cézanne, and Pissarro, had held an exhibition in the studio of a Paris photographer. Visitors had openly expressed their bewilderment and dismay over paintings made of thousands of tiny dots and strokes that blurred together into blotches and crude images, paintings with colors extravagantly bright and overlapping impossibly. One of Monet's paintings, a painting of the sun rising through the fog over the harbor at Le Havre and shimmering in the water, had attracted the attention of one journalist, who took the name of the painting, *Impression: Sunrise,* to label the whole group and their style. The name had stuck. Their work was impressionism.

By the ceaseless turning of the spheres, the darkness of the sky turned into dawn. From the combined light of the trillions of stars of the galaxy whose spiral filled most of the sky, the waters had flickered all through the night, and now as the sun appeared, the flickering turned into a brilliant glare dancing with even brighter sparkles. Cresting waves formed thousands of lenses that focused the light into gleams, while water spraying into the air formed prisms that broke the light apart into its colors. Yet these flickers of color were tiny amid the vast silver of the ocean. The silver was unbroken by any land anywhere upon the planet. Every sunrise at every point on the planet was an ocean sunrise, a brewing of water and light. Clouds gave the light many flavors. Like the water whose scintillation revealed its constant motion, the clouds too were constantly flowing, changing position, shape, and thickness, making each sunrise snowflake-unique. The sea, the air, the clouds—everything upon this planet flowed as steadily as the stars flowing through the spiral in the sky.

Eager to capture the exact flavor of the moment, Monet studied the way the light gleamed up from the water, how it carried the colors and reflections and shadows of the lilies, how it revealed the subtle flowing of the pond. The water of the pond was channeled from the Epte River and exited out a further channel and back into the river; the Epte River soon flowed into the Seine, which, a hundred miles downstream at Le Havre, flowed into the sea, joining the vast blue flowing that covered two-thirds of the planet. One day the water Monet was invoking on his canvas would rise into the sky and, as clouds, drift back over the land and fall onto the green hills of France and flow again down the Epte, through this pond, and flow down the Seine it had traveled a thousand times before. Monet was capturing one tiny part of the ceaseless flowing of huge forces, the flowing of water, air, clouds, sunlight, and land; the turning of the planet; the flowing of the planet through space; the flowing shapes of the four seasons; and, yes, the flowing shapes of life through four billion years.

By the ceaseless turning of the spheres, the darkness of the sky quickly turned into dawn. There was little warning of the nearness of the sun, for the planet held little atmosphere to spread the light. Suddenly the sun was there, small because distant, yet bright enough for its light to enter the crystals of the ice and come out gleaming a vague blue. The sun was mirrored up from plains of ice cracked into chasms. On the horizon was a solitary mountain of ice, and here the sunrise first became visible as a glowing point afloat in the sky, a point that slowly expanded downward into a pyramid, taller and taller, glowing blue.

No painter before him had taken as much interest in rendering the fleeting moods of the day. Because one painting might not communicate which colors or glows in a scene were lasting and which were fleeting, he often painted a whole series of paintings of the same scene, each at a different time of day. He would paint the scene from the same angle, from the exact same spot, working on one canvas for twenty or thirty minutes until the light changed, and then he put that canvas aside until the next day and worked on another canvas of the same scene for another thirty minutes. The objects to be painted were almost irrelevant as long as they revealed the changing behavior of light. He had painted a series of the Houses of Parliament reflecting up from the Thames, a series of Venice, and a series of the Rouen Cathedral. Yet nature's architecture around his house was quite sufficient in subjects, and he spent years doing dozens of paintings of poplars, poppy fields, haystacks, his garden, and most of all the pond, which he painted over and over and over again to capture every slight variation.

By the ceaseless turning of the spheres, the darkness of the sky quickly turned into dawn. There was little warning of the nearness of the sun, for the planet held little atmosphere to spread the light. Suddenly the sun was there, small because distant, yet bright enough for its light to enter the crystals of the ice and come out gleaming a vague blue. The sun was mirrored up from plains of ice cracked into chasms. On the horizon were two mountains of ice, and here the sunrise first became visible as two glowing points afloat in the sky, points that slowly expanded downward into two pyramids, taller and taller, glowing blue.

Pausing for a moment, Monet reached into his jacket pocket and lifted out a handful of cherries, which he had picked yesterday morning from one of the Japanese cherry trees in his garden. As he studied the dawn, he placed into his mouth one by one the cherries as round and red as the rising sun. Many times he had captured on canvas the green, brown, pink, and red flavors of the sunlight shining on those cherry trees; as his tongue squeezed open the cherries, he tasted through a different sense the flavor of that sunlight. Monet's canvas had not been the only surface to capture that sunlight. The green leaves of the cherry trees had also captured some of it. The captured sunlight had flowed throughout the trees and energized the raising of water and nutrients through roots and up trunks and branches, energized the inhaling and exhaling of gasses, energized the binding of atoms into molecules as learned as DNA, energized the thousand activities inside cells, energized the multiplication of cells into the careful shape and color of a cherry. Each cherry was packed with little pieces of the sun, and as Monet swallowed, that sunlight flowed into him, joining the streams of sunlight already flowing through him, joining activities more sophisticated than those within plants. From his heart the sun poured warm red rays throughout his body. In his muscles the sun—in one small eddy of the gravitational currents on which the solar system flowed—pulled against the gravity of Earth to keep his body risen. Within his brain when he awoke this morning the sun had dawned as consciousness, and now his brain glowed with the shapes and colors of the landscape. Through his eyes the sun was seeing itself rising over that landscape, and through his hand the sun was painting a self-portrait.

By the ceaseless turning of the spheres, the darkness of the sky turned into dawn.

Amid thousands of planets where dawn only stirred up the dust or clouds, on this planet the growing light stirred cells into activity. For a long time upon this planet the dawn had been grey, its two suns glowing over plains of grey dust and rock. When volcanoes had pumped this greyness into the sky, the sunlight often barely penetrated to the ground. Gradually the clouds turned from grey to white, and the dawns gleamed upon oceans and lakes. The waters became tinged with green, and eventually that green overflowed the water and spread across the land and arose into millions of shapes.

Now the dawn was green, not just because the sunlight flowed onto the plants and illuminated them from the outside, but because the sunlight was also entering the plants and becoming another kind of burning, a green fire. During the night this green fire had ebbed, and many of the leaves had drooped or folded. As the two suns rose in the sky they also rose from the ground as billions of round and oval and odd-shaped leaves, as green suns.

The canvas was almost full. Light glowed from the narrow strip of sky at the top of the painting, and it glowed again from the water. Bright colors glowed from the grass and flowers and bushes and trees on the bank, from the lilies and fallen leaves floating on the water, and glowed again from their reflections. In less than an hour dozens of forms of life had emerged from the blankness of the canvas, a reenactment of that greater blooming through which all these forms had arisen. Four billion years ago the surface of Earth had been as blank as this canvas had started out, only plains of grey dust and rocks and craters. But as sunrise after sunrise swept over those plains, sunrises adding into hundreds of millions of years, their energy stirred the dust from the ground and gave it shape and color. The sunlight was a luminous brush that painted the planet, filling every inch of its surface with life. Every morning that brush had stroked across the land and water and added to and altered the living forms and colors, tiny changes that flowed into massive changes. That brush began by drawing a single cell, and it copied that cell over and over until the oceans were tinged with green. It spread that green over the land and mixed it with brown streaks to thicken the green into the air, each leaf a brush stroke, blending to make whole continents green. In the water that brush painted fish, and then it drew legs on them and swarmed them over the land. It painted some of those animals into birds and colored the skies with them. It painted insects to accent the land. It created many basic forms of life and then painted millions of variations on each form, painting each variation over and over again, some of them billions of times every hour. It painted creatures in fantastic detail, inside and out. It gave them every color and mixture of colors. It dipped into the blueness of the sea and fixed that blueness onto the skin of whales. It painted bees as yellow as the biologized suns they harvested. It painted peacocks as walking rainbows to dazzle genes into renewal. It painted on ravens the blackness of space. It painted millions of kinds of eyes to flick open every morning, eyes through which the flowing light was focused into brains, into those grey canvases where all the forms and colors of life were painted again, where the same shape and color took on a different appearance to each species. And then, four billion years into its work, the light painted the shape of a human standing beside a pond, a painter painting the light.

By the ceaseless turning of Earth, the darkness of the sky turned into dawn. The light swept over mountains and plains, rivers and oceans and deserts, all the landscapes it had been illuminating for billions of years. It swept across the more recent living landscapes, the prairies and forests and marshes, setting them glowing and flowing. And then, as the light flowed into one river valley, it found standing beside a pond a package of trillions of cells, throughout which sunlight pulsed. From the vagueness of night the dawn sketched out the lines of a face, the thicket of a beard, the colors and textures of clothes. Into two eyes the sunlight poured, and after circling through a brain, that sunlight came out again through a hand and affixed itself to a canvas. The light pulsing in those cells powered each shift of the eyes, each thought-bolt in the brain, each stroke of the brush. Those trillions of cells were portraying the behavior of light, not just with paint, but far more profoundly by their very act of painting. Monet is the word for how light behaves after fourteen billion years of cosmic evolution.

The thing he enjoyed most about painting was how it focused the mind, how it pushed aside mental chatter and left your attention filled and vibrant with the present moment, the immediate scene. It was a form of meditation through which the most subtle aspects of the world revealed their majesty.

Amid hills and rivers and wind that had always been numb, amid animals for whom the dawn only clocked their daily activities, Monet was a pure, bright consciousness absorbing the universe.

He reached out and placed one last stroke of yellow on the canvas, which was also a finishing touch on a much greater canvas.

The ceaseless turning of the spheres.

Within a single galaxy there might be billions of planets rolling steadily onward, and throughout the billions of galaxies in the universe, the number of planets was immense. Since each of those planets was constantly enacting a sunrise somewhere upon it, the universe was crowded with sunrise. The sunrises happened with great variety. The suns were large or small; near or far; single or double or triple; expanding, contracting, pulsing, or stable; colored red, blue, white, or yellow. The suns rose over planets large or small, solid or gaseous, planets of every color and mixture of colors. They rose through nebulas still turning into new planets, and they rose over planets ten billion years old. They rose fast or they rose only one degree per hour. They rose as savage heat or as specks of light barely larger than other stars. They rose over planets whose nakedness to space made the sunrise a sudden burst of light, and they rose through atmospheres that diffused the light into a premonition and then a gentle blooming. They rose through atmospheres of every thickness and velocity and color, and their light was shaded differently by every one, and shaded differently from minute to minute as each atmosphere flowed. They rose over landscapes of nothing but craters, and slowly climbed into them. They rose over fields of scattered rock, of sand dunes, of dust. They rose over mountain ranges, and they crept into canyons until they glowed from the river at the bottom. They rose gleaming over oceans, bays, and lakes. Their fire collided with the fire pouring from volcanoes. They rose through snowfalls, and over planets of solid ice. They set off momentary rainbows, and stirred the ancient yet always changing colors of gas planets. And on those millions of planets whose ceaseless turnings had turned them into life, the rising sun printed cells. The suns rose over oceans swarming with fish; over green plains; over forests and jungles; over ponds brimming with reeds, algae, and lilies; over skies cloudy with birds; over megaliths and ceremonies; over cities gleaming with millions of suns of their own.

And all of these sunrises, all these innumerable turnings of darkness into light, were but tiny flickers of a much vaster sunrise, a greater coming of energy and activity. This sunrise ended a night that had lasted forever, a massive night totally dark and empty, containing not one glow of light, not one ripple of energy, not one atom, no activity or matter of any kind, only infinite emptiness. Suddenly that darkness had been shattered by a blast of light. The light began as a single point of infinite intensity, and at once it began to swell into a sphere of brilliant light, ceaselessly growing. This sphere contained all the energy the universe would ever have and everything the universe would ever become. This was the dawning of a cosmos, a sunrise called the Big Bang, a dawn that would continue to expand until it was so vast that it broke into trillions upon trillions of tinier suns which would bathe the universe with day.

Those trillions of separate suns, each a brush stroke of bright color emerging from a canvas that had been black, blended together to form larger patterns, such as star clusters, the various shapes of galaxies, and clusters of galaxies. Yet all of these shapes were but details of one larger image, an image billions of light-years wide. All of those trillions of dots of light combined to form an impressionistic portrait of a sunrise.

The energy set flowing by that cosmic sunrise would take on many forms. In the first instants of the Big Bang that energy dressed itself up as matter. It became the massive flowing of light from stars. It became the momentum that kept planets turning for billions of years and kept them reenacting over and over on a tiny scale the dawn from which the universe began. It became the ceaseless flowing of water, air, and land. It became the intricate flowings in cells and all the activities and feelings of life. It became Claude Monet awakening and rising to meet the rising sun. It enhanced itself into the energy pulsing through his brain and moving his eyes and hands. Through his hands it became a further kind of reenactment of sunrise. In such brains the cosmic dawn reached a further illumination, becoming brilliant enough for it to see itself, becoming intense enough for light to fall in love with the changing moods of light on the surface of a pond or in the flow of a galaxy, becoming penetrating enough to reveal the glory of that roaring sunrise expanding ever father into space.

Endnotes

1 Quoted in Harlow Shapley, *Through Rugged Ways to the Stars: The Reminiscences of as Astronomer* (New York: Charles Scribner's Sons, 1969), 78.

2 Harlow Shapley, *The View From a Distant Star: Man's Future in the Universe* (New York: Dell, 1967), back cover.

3 Ibid., 79.

4 Ibid., 37.

5 Harlow Shapley, *Beyond the Observatory* (New York: Charles Scribner's Sons, 1967), 171.

6 Ibid., 151.

7 Shapley, *The View From a Distant Star*, 77.

8 Ibid., 36.

9 Ibid., 38.

10 Ibid., 160.

11 Ibid., 22.

12 Ibid., 77.

13 Shapley, *Beyond the Observatory*, 100.

14 Shapley, *Through Rugged Ways to the Stars*, 57.

15 Ibid., 57.

16 Quoted in Tony Ortega, "Red Scare at Harvard," *Astronomy*, January 2002, 44.

17 *History of Hickory, Polk, Cedar, Dade, and Barton Counties, Missouri*, (1889), 942.

18 Shapley, *Through Rugged Ways to the Stars*, 5.

19 Lamar, Missouri, newspaper, August, 1914. Quoted, without further reference, on website of local historian Stephen Crouch.

20 Shapley, *Through Rugged Ways to the Stars*, 42.

21 Quoted in Samuel Keller, "An Infidel Experiment," *St. Louis Post-Dispatch*, May 2, 1885.

22 Quoted in J. P. Moore, *This Strange Town—Liberal, Missouri, A History of the Early Years: 1880-1910* (Liberal, Missouri: The Liberal News: 1963), 34.

23 Ibid., 35.

24 Harlow Shapley, "Stars, Ethics, and Survival," *Science Ponders Religion* (New York: Appleton-Century-Crofts: 1960), 11-12.

25 Shapley, *Through Rugged Ways to the Stars*, 20.

26 Ibid., 66.

27 Shapley, *The View From a Distant Star*, 21.

28 Ibid., 161.

29 Shapley, *Through Rugged Ways to the Stars*, 18.

30 Harlow Shapley, *Of Stars and Men: The Human Response to an Expanding Universe* (New York: Washington Square Press, 1959), 22.

31 Shapley, *The View From a Distant Star*, 82-3.

32 Quoted in George Johnson, *Miss Leavitt's Stars* (New York: W. W. Norton, 2005), p. 67.

33 Cecilia Payne-Gaposchkin, *Cecilia Payne-Gaposchkin: An Autobiography and Other Recollections* (Cambridge, England: Cambridge University Press, 1984), p. 150.

34 Harlow Shapley, *Through Rugged Ways to the Stars: The Reminiscences of an Astronomer* (New York: Scribner's, 1969), p. 93.

35 Quoted in Eugeni Ivanovich Riabchikov, *Russians in Space* (Garden City, N. Y.: Doubleday, 1971), p. 92.

36 Quoted in Maynard Solomon, *Beethoven* (New York: Simon and Schuster Macmillan, 1998), p. 153.

37 George Hale, "Some Personal Reflections," unpublished.

38 George Hale, "Work for the Amateur Astronomer," *Publications of the Astronomical Society of the Pacific*, volume XL (1928), 285-302.

39 Robert Ball, *The Story of the Heavens* (London; Cassell, 1885).

40 Theodore Dreiser, *The Financier* (New York: The World Publishing Co, 1946), 468.

41 Ibid., 3.

42 Ibid., 4-5.

43 Ibid., 226.

44 Theodore Dreiser, *The Titan* (New York: New American Library, 1984), 337-8.

45 Theodore Dreiser, *The Financier*, 364.

About the Author

For nearly thirty years Don Lago's explorations of nature and science have appeared in numerous magazines, including *Astronomy, Sky & Telescope, Orion, Earth, Air and Space Smithsonian, Science Digest, Antioch Review,* and *Michigan Quarterly Review.* He has won several awards, including the 2009 Editor's Prize from *Isotope.* He is the author of *On the Viking Trail: Travels in Scandinavian America* (University of Iowa Press, 2004). He was the science book reviewer for *Saturday Review.* He lives in a cabin in the pine forest in Flagstaff, Arizona. He also explores deep time by hiking and kayaking in the Grand Canyon.

Photo by Tom Bean

www.ingramcontent.com/pod-product-compliance
Lightning Source LLC
Chambersburg PA
CBHW071701210326
41597CB00017B/2274